Caledonian Structures
in Britain

THE GEOLOGICAL CONSERVATION REVIEW SERIES

The comparatively small land area of Great Britain contains an unrivalled sequence of rocks, mineral and fossil deposits, and a variety of landforms that span much of the earth's long history. Well-documented ancient volcanic episodes, famous fossil sites, and sedimentary rock sections used internationally as comparative standards, have given these islands an importance out of all proportion to their size. These long sequences of strata and their organic and inorganic contents, have been studied by generations of leading geologists thus giving Britain a unique status in the development of the science. Many of the divisions of geological time used throughout the world are named after British sites or areas, for instance the Cambrian, Ordovician and Devonian systems, the Ludlow Series and the Kimmeridgian and Portlandian stages.

The Geological Conservation Review (GCR) was initiated by the Nature Conservancy Council in 1977 to assess, document, and ultimately publish accounts of the most important parts of this rich heritage. Since 1991 the task of publication has been assumed by the Joint Nature Conservation Committee on behalf of the three country conservation agencies, English Nature, Scottish Natural Heritage and the Countryside Council for Wales. The GCR series of volumes will review the current state of knowledge of the key earth-science sites in Great Britain and provide a firm basis on which site conservation can be founded in years to come. Each GCR volume will describe and assess networks of sites in the context of a portion of the geological column, or a geological, palaeontological, or mineralogical topic. The full series of approximately 50 volumes will be published by the year 2000.

Within each individual volume, every GCR locality is described in detail in a self-contained account, consisting of highlights (a précis of the special interest of the site), an introduction (with a concise history of previous work), a description, an interpretation (assessing the fundamentals of the site's scientific interest and importance), and a conclusion (written in simpler terms for the non-specialist). Each site report is a justification of a particular scientific interest at a locality, of its importance in a British or international setting, and ultimately of its worthiness for conservation.

The aim of the Geological Conservation Review series is to provide a public record of the features of interest in sites being considered for notification as Sites of Special Scientific Interest (SSSIs). It is written to the highest scientific standards but in such a way that the assessment and conservation value of the site is clear. It is a public statement of the value set on our geological and geomorphological heritage by the earth-science community which has participated in its production, and it will be used by the Joint Nature Conservation Committee, English Nature, the Countryside Council for Wales, and Scottish Natural Heritage in carrying out their conservation functions.

All the sites in this volume have been proposed for notification as SSSIs, the final decision to notify or renotify lies with the governing Councils of the appropriate country conservation agency.

Information about the GCR publication programme may be obtained from:

Earth Science Branch,
Joint Nature Conservation Committee,
Monkstone House,
City Road,
Peterborough PE1 1JY.

Titles in the series

Caledonian Structures in Britain

South of the Midland Valley

Edited by
J. E. Treagus

Department of Geology,
The University, Manchester

GCR editor: W. A. Wimbledon

CHAPMAN & HALL

London · Glasgow · New York · Tokyo · Melbourne · Madras

Published by Chapman & Hall, 2–6 Boundary Row, London SE1 8HN

Chapman & Hall, 2–6 Boundary Row, London SE1 8HN, UK

Chapman & Hall, 29 West 35th Street, New York NY10001, USA

Chapman & Hall Japan, Thomson Publishing Japan, Hirakawacho Nemoto Building, 6F, 1–7–11 Hirakawa-cho, Chiyoda-ku, Tokyo 102, Japan

Chapman & Hall Australia, Thomas Nelson Australia, 102 Dodds Street, South Melbourne, Victoria 3205, Australia

Chapman & Hall India, R. Seshadri, 32 Second Main Road, CIT East, Madras 600 035, India

First edition 1992

© 1992 Joint Nature Conservation Committee

Typeset in 10/12 Garamond by Columns Design & Production Services Ltd, Reading
Printed in Great Britain at the University Press, Cambridge

ISBN 0 412 47560 X

A catalogue record for this book is available from the British Library

Library of Congress Cataloging-in-Publication data available

Contents

Contents

Contributors

Dr D. E. B. Bates Institute of Earth Studies, University College of Wales, Aberystwyth, Dyfed, SY23 3DB.

Dr A. M. Bell 143 Graham St, Penrith, Cumbria, CA11 9LG.

Dr A. H. Cooper British Geological Survey, Windsor Court, Windsor Terrace, Newcastle upon Tyne, NE2 4HE.

Dr W. R. Fitches Institute of Earth Studies, University College of Wales, Aberystwyth, Dyfed, SY23 3DB.

Dr F. Moseley 89 Cambridge Rd, Birmingham, B13 9UG.

Dr R. Nicholson Department of Geology, The University, Manchester, M13 9PL.

Dr D. E. Roberts Department of Earth Sciences, Staffordshire Polytechnic, College Rd, Stoke-on-Trent, ST4 2DE.

Dr R. Scott Department of Earth Sciences, Downing St, The University, Cambridge, CB2 3EQ.

Dr J. E. Treagus Department of Geology, The University, Manchester, M13 9PL.

Dr B. C. Webb 1 Grassfield Cottages, Nenthead, Nr Alston, Cumbria, CA9 3EQ.

Dr N. H. Woodcock Department of Earth Sciences, Downing St, The University, Cambridge, CB2 3EQ.

Acknowledgements

Work on this volume was initiated by the Nature Conservancy Council and has been seen to completion by the Joint Nature Conservation Committee (JNCC), acting on behalf of the three country conservation agencies, English Nature, Scottish Natural Heritage and the Countryside Council for Wales.

Each site description bears the name of its author, but we would like to acknowledge the many colleagues who gave help and advice. In particular, we would like to thank Professor P. Stringer, Drs A. E. S. Kemp, J. A. McCurry, N. J. Soper and B. C. Webb for their assistance. Dr L. E. Richards of the Nature Conservancy Council was responsible for the volume in its early stages. Drs J. E. Treagus, F. Moseley and W. R. Fitches initially selected and edited the accounts of the sites of the Southern Uplands, Lake District, and Wales, respectively. Dr K. Fraser was employed for six months in 1988 to word-process and to assist Dr Treagus in editing the first draft of the volume. Dr P. H. Banham, Dr N. H. Woodcock and Professor J. L. Knill scrutinized the subsequent text, and some later GCR editing was undertaken by Dr W. A. Wimbledon.

Thanks also go to the GCR publication production team: Dr D. O'Halloran (Project Manager); Valerie Wyld (Sub-editor); Nicholas D. W. Davey (Cartographic editor); and Caroline Mee (Administrative assistant). Computerized cartographical illustrations were produced by Silhouette (Peterborough).

Acccess to the Countryside

This volume is not intended for use as a field guide. The description or mention of any site should not be taken as an indication that access to a site is open or that a right of way exists. Most sites described are in private ownership, and their inclusion herein is solely for the purpose of justifying their conservation. Their description or appearance on a map in this work should in no way be construed as an invitation to visit. Prior consent for visits should always be obtained from the landowner and/or occupier.

Information on conservation matters, including site ownership, relating to Sites of Special Scientific Interest (SSSIs) or National Nature Reserves (NNRs), in particular counties or districts, may be obtained from the relevant country conservation agency headquarters listed below:

Scottish Natural Heritage,
12 Hope Terrace,
Edinburgh EH9 2AS.

Countryside Council for Wales,
Plas Penrhos,
Ffordd Penrhos,
Bangor,
Gwynedd LL57 2LQ.

English Nature,
Northminster House,
Peterborough PE1 1UA.

Preface

This volume deals with those sites selected as part of the Geological Conservation Review (GCR) within the southern British part of the Caledonides, that is, the paratectonic Caledonides – a Caledonian terrane without strong and pervasive deformation and metamorphism, such as occurred further north. This orogenic belt formed by long and complex processes of earth movements between 500 and 380 million years before the present (?late Cambrian to mid-Devonian times), and has been classic ground for geologists for two hundred years. It is perhaps no accident that James Hutton in 1795 chose to illustrate his geostrophic cycle (and unconformity) with three visually explicit examples of the deformation wrought on Lower Palaeozoic rocks by Caledonian events.

The former Caledonian mountain chain, which can be seen today in fragmented pieces in Scandinavia, Britain and Ireland, and North America, was ultimately the result of the collision of two continental plates and the closure of a former ocean, Iapetus. Some of these fragments, including those in Scandinavia, southern Britain, and the Republic of Ireland and the Maritime Provinces of Canada, are thought to have lain on the south side of the ocean before collision: the rest of North America, northern Ireland, and Scotland are thought to have lain north of the former Iapetus. The width of the late Precambrian to Early Palaeozoic ocean, at various stages before its closure, has been greatly debated (McKerrow and Cocks, 1976; Phillips *et al.*, 1976). Much concerning its formation, its narrowing and destruction, and the tectonic (and plate tectonic) consequences of these events has yet to be elucidated, but it is clear that the mountain chain that formed by mid-Devonian times was once continuous across what has become the Atlantic area, and that deformation phases which affected the rocks on the western side of the Atlantic are comparable with those of the Taconic and Acadian orogenies in the Appalachians (Bailey, 1929). Before the opening of the north Atlantic, around 60 million years before the present, the eroded Caledonides with their characteristic NE–SW tectonic grain, stretched for some 5000 kilometres, from the Arctic to the southern United States.

The orthotectonic Caledonides of Britain, that is those areas affected by metamorphism and tectonism north of the Highland Boundary Fault, will be dealt with in subsequent volumes of the GCR series. This volume describes key sites demonstrating Caledonian tectonism in Wales, the Lake District, and in Scotland, south of the Midland Valley. The first two areas lay, in pre-collision times, on the south side of Iapetus, Wales being the site of the deposition of a great thickness of Early Palaeozoic sediments and volcanics in a marginal basin. Of course, the Welsh Basin is even more famous for containing the type areas and type localities for the Early Palaeozoic Cambrian, Ordovician and Silurian systems. The Caledonian structures in these rocks show particularly the influence of structures in the (Precambrian) basement. The Lake District has been interpreted as being the setting of an Early Palaeozoic island arc, in some interpretations lying south of a subduction zone in which the south-eastwards-moving oceanic floor of Iapetus was being destroyed.

In Early Palaeozoic times, Scotland, including the areas and sites in this volume in southern Scotland, lay on the opposite side of Iapetus to those in Wales and the Lake District above the complementary subduction zone. Rocks in the Southern Uplands have been interpreted as the product of an accretionary prism, that is as wedges of ocean floor pushed, thrust, and welded (accreted) by plate tectonic movements on to the north-western continental margin of Iapetus. These rocks include deep-sea sediments mixed with slivers of ocean crust on which they had been deposited during the Ordovician and possibly the Silurian Period, all carried on to the continental margin, lying to the north of the putative subduction zone, as they were 'scraped off' the back of the subducting ocean plate.

The Southern Uplands have been a proving ground for tectonic models and for testing the constraints imposed by the vitally important graptolite biostratigraphy. One model, for instance, suggests that subduction and deformation may have ceased on its northern margin by the early Silurian Period. In this view, the early Devonian culmination of the Caledonian deformation is really confined to the folds and cleavage of the Lake District and Wales, although the subsequent sinistral fault movements throughout the British area provide final, unifying evidence of Iapetus' closure.

The present site descriptions were initiated in 1983, building to some extent on the small coverage of existing SSSIs. The Southern Uplands sites were mostly visited and described in 1983/4, with some updating in 1986, whereas the Lake District and Welsh site descriptions were not completed until 1988. Contributors were asked to employ the standard Geological Conservation Review criteria, that is to identify sites of national importance, to describe their features, and detail the scientific justification for GCR selection and ultimately SSSI notification. However, a slightly different approach was required in assessing structural sites than would be required when, for instance, selecting more conventional stratigraphical or palaeontological sites; the guidelines followed were that localities should be selected for structural features which best illustrated Caledonian deformation, but in three distinct subareas. In effect, this meant scrutiny of the literature and canvassing of expert opinion in identifying all sites which were known to exhibit important structural features to advantage and, to some extent, the seeking out of sites which might display a particular structural characteristic. From this preliminary list it was necessary to select those localities which best illustrated the typical features of the various phases of deformation, as well as the principal variations and exceptions. Many potential sites had to be excluded because they duplicated features seen elsewhere; and it has not been possible to illustrate some aspects of deformation as no appropriate site was known.

This last point raises the matter of the great burst of geological research activity there has been in the areas described, since the site descriptions were written. All three areas have not only undergone considerable scrutiny by academic researchers in the last three years, but have also been locally subject to very detailed attention from the British Geological Survey. This has revealed many new potential sites which illustrate known features of Caledonian deformation or features that have acquired a new significance as research has progressed. In this latter category are sites which might better illustrate the timing of Caledonian events (for example, as a result of the dating of igneous bodies, cleavage), details of internal processes (for example, fracture systems, cleavage development, shear criteria, strain variations) and evidence of external processes (for example, plate movement, major fault displacements). It will be noted that sites from the Midland Valley have been excluded from this volume. At the time of its preparation there was very little information available concerning Caledonian structure, but it was clear that this was to become an area of significance in the 'jigsaw' of Caledonian evolution. Being so intimately involved in the stratigraphy, sedimentology, and igneous history, the structure of the area will be covered in later relevant GCR volumes.

W. A. Wimbledon

Chapter 1

Caledonian structures

INTRODUCTION

J. E. Treagus

The purpose of this volume is to describe and discuss the selected Geological Conservation Review sites which demonstrate structures of Caledonian age (Cambrian to early Devonian), south of the Southern Upland Fault. The sites, of national importance, have been selected to illustrate all the principal features of the Caledonian Orogeny in Britain (Scotland, England and Wales).

The volume has been divided into three sections; the Southern Uplands, the Lake District and Wales, for both geographical and geological reasons. Each of these sections is introduced and an outline of its structural features given, putting the sites into a Caledonian context. The purpose of the following paragraphs is to introduce the principal features of the Caledonian Orogeny in Britain south of the Southern Uplands Fault, so that the three-component sections can be seen in the context both of the British area and of the wider setting of the Caledonian–Appalachian Orogenic Belt.

THE CALEDONIAN OROGENIC BELT

The Caledonian–Appalachian Orogen can be traced (pre-Atlantic drift), for some 7500 km south-west to north-east, from south-eastern USA through the British Isles to Scandinavia, Greenland, and Ny Friesland (Figure 1.1). It is generally accepted, after the work of Wilson (1966) and Dewey (1969), that sedimentation and igneous activity took place at, or near, the margins of an ocean (the Iapetus) that separated the Laurentian and Gondwanaland plates, over a period from the Precambrian through the early Palaeozoic. From studies of fauna, sedimentary history, igneous activity, structural and metamorphic evolution, and palaeomagnetism on its two sides, it is considered that deformation of sediments and volcanics, resulting from the episodic closure of the Iapetus Ocean, took place through the early Palaeozoic to culminate in continental collision during the early Devonian.

Since the initial plate tectonic model for this orogen (Dewey, 1969; see Figure 1.2), many variations and refinements have been proposed. Subduction and obduction of the Iapetus oceanic crust is regarded as having occurred on both sides of the ocean at various times, with attendant accretion and deformation. The opening and closing of marginal basins, the collisions of volcanic arcs and microcontinents, as well as the final continent to continent collision and suturing, have all been used to explain various deformation events (Barker and Gayer, 1985). It has long been recognized that various regions within the orogen have had very different histories and different deformation timings (see Figure 1.1), but recently emphasis has been placed on the differences between smaller areas (terranes) with distinct geological histories within these regions. In particular, the work of Williams and Hatcher (1982) in the Appalachians, has led to the recognition that thrusting and strike-slip movements may be responsible for the very distinct geological histories of many of these terranes (see Barker and Gayer, 1985).

THE BRITISH CALEDONIDES

Scottish Highlands

In the British Isles (Figure 1.3) two groups of contrasting terranes are of long standing: the Scottish Highland Terranes (and Irish equivalents) with their early, 590–480 Ma Caledonian (= Grampian) deformation and metamorphism, and the Southern Upland, Lake District, and Welsh Terranes (and their Irish equivalents) dominated by late or end-Caledonian (= Acadian) deformation and low-grade metamorphism, dated around 400 Ma. The first of these areas itself has a complex history, but culminated with the Grampian Orogeny imposed on rocks ranging in age from Archaean basement (Lewisian) through the Proterozoic (Moine) to the Late Proterozoic to early Cambrian (Dalradian) cover. The principal deformation events took place before 490 Ma, resulting in polyphase folding, thrusting, and regional low- to high-grade metamorphism. Subsequent folding in this area has not been precisely dated, but it pre-dates intrusion of Upper Silurian to Lower Devonian granites and faulting on the Great Glen Fault system. Some of this late folding, and certainly the faulting, must correlate in time with the late-Caledonian deformation of the terranes to be considered below.

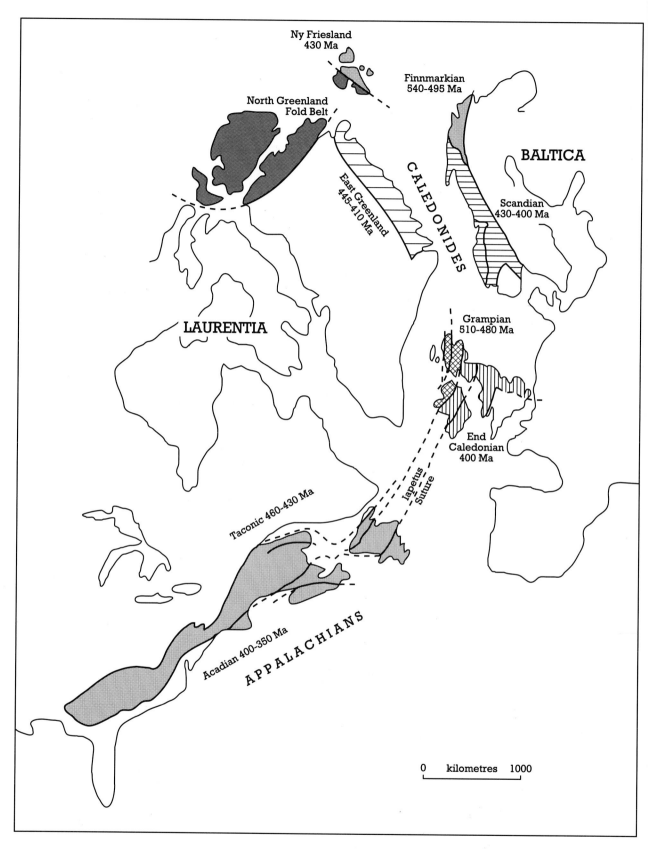

Figure 1.1 Regions of the Caledonian–Appalachian Orogen in their pre-Mesozoic drift configurations, showing ages of principal deformation events (after Barker and Gayer, 1985).

A)

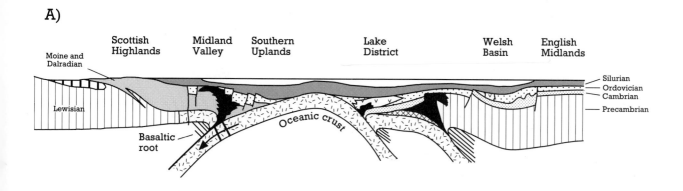

B)

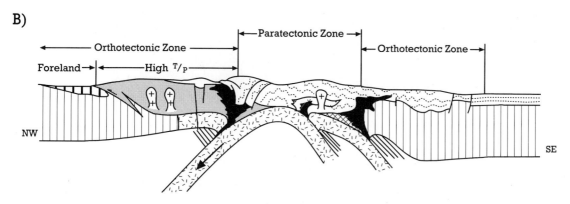

Figure 1.2 Schematic cross-sections of the Caledonides, after Dewey (1969, figure 2E and F). (A) represents Iapetus during the Silurian. (B) shows the situation after collision in the early Devonian, with ornament indicating fold style in Lower Palaeozoic rocks. Black areas represent volcanics and intrusions of the Ballantrae Complex (NW) and Gwna Group of Anglesey (SE); Vs represent Upper Ordovician volcanics of the Lake District and Wales.

Midland Valley

Between the Scottish Highlands and the Southern Uplands lies the enigmatic terrane of the Midland Valley. Apart from some indirect evidence here of a granulite basement, the oldest exposed rocks comprise the Lower Ordovician Ballantrae Complex, commonly interpreted as obducted Iapetus Ocean crust, with a complex and little-understood terrane history. These rocks are unconformably overlain by Middle Ordovician to Middle Silurian sediments which contrast strongly in their sedimentology and structural and metamorphic state with the rocks of the Southern Uplands. In the south-east of the Midland Valley, they are very weakly folded and appear to be conformable with the Lower Devonian, although folding preceded Gedinnian Series deposition in the north-east in the Pentland Hills. The Middle Devonian is missing and the Upper Devonian is strongly unconformable on older rocks, with evidence, in this interval, of locally strong folding and faulting.

Another element of the Midland Valley Terrane is a narrow zone of possibly ophiolitic rocks which parallels the Highland Boundary Fault: this has small areas of Arenig Series, Middle Ordovician and Upper Ordovician sediments, each apparently with a distinct structural history.

The Midland Valley has not only a contrasting history compared with terranes to its north-west and south-east, but also a lack of features, sedimentological, magmatic, and structural, that might support continuity. This has led recent workers (for example, Bluck, 1986; Hutton, 1987) to favour the idea that the Midland Valley owes its present position largely to strike-slip movements on its two boundary faults, in preference to previous attempts to integrate it directly into Dewey's (1969; Figure 1.2) destructive plate margin. The history of the Midland Valley is very important to the understanding of Caledonian evolution of the British area, but because so much work is in progress, its consideration has been excluded from the current volume.

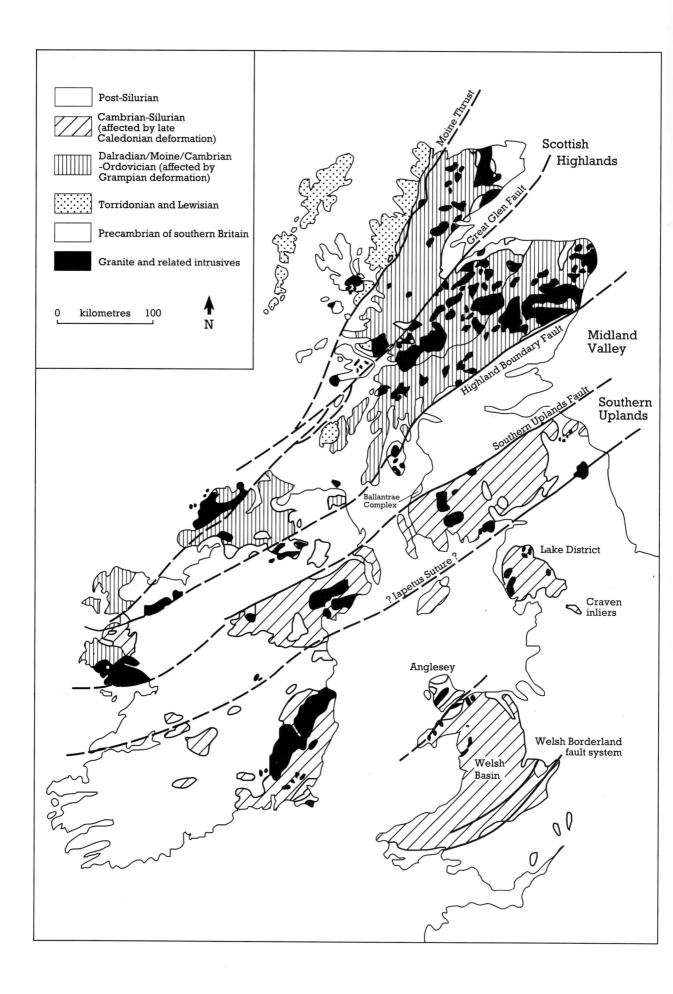

Southern Uplands

The deformation events that are the subject of this volume are generally construed to be the result of the closing of the Iapetus between the mid-Ordovician Period and the early Devonian, associated with marginally directed subduction zones. The models for this closure are largely based on Dewey (1969) (see Figure 1.2) and there have been many suggestions as to how both large-scale and small-scale structures may be related to these models. On the north-western margin, the arguments for NW-directed subduction in the Southern Uplands have been strengthened by the stratigraphical, sedimentological, and large-scale structural evidence. The distinctive stratigraphical and structural arrangement of these rocks has been used to argue, very persuasively, (McKerrow *et al.*, 1977), for accretion above a descending oceanic plate. According to the accretionary prism model, deformation in the Southern Uplands, unlike that in the Lake District and Wales, would be expected to have developed throughout late Ordovician and Silurian times, perhaps culminating with the development of the finite cleavage in the early Devonian. Thus the upright SE-verging and steep reverse faults have been interpreted as original flat-lying, ocean-verging and NW-dipping thrusts respectively, which have been rotated into their present steep attitudes in the accretion process. The cleavage, which cuts across the folds, has been attributed to the late-stage closure of the Iapetus. However, there is still much debate concerning the detailed relationship of the structures observed to the evolution of the supposed accretionary prism (a wedge-shaped pile of deformed rocks) above the subducting plate. Indeed, recently, doubts have been cast upon the reality of the accretionary prism model, particularly as applied to the Silurian rocks (Hutton and Murphy, 1987).

Lake District

On the south-eastern margin, the argument for SE-directed subduction (Figure 1.2) rests largely on the presence of arc volcanism in the Lake District during the Llandeilo or Caradoc epochs. Structural

Figure 1.3 Simplified map of the British Caledonides modified from Leake *et al.* (1983).

arguments have focused on the nature and significance of the pre- and post-Borrowdale Volcanic Group unconformities, but particular features that could be related directly to subduction have not been identified. Folding in the Skiddaw Group has recently been reinterpreted as being the product of slumping (Webb and Cooper, 1988), and this may prove important with regard to the arguments for the timing of subduction and the topography of the margin. Apart from tilting and block faulting due to volcanotectonic activity (Branney and Soper, 1988), the dominant deformation in the Lake District is regarded as essentially a single event that resulted in folding, cleavage, and greenschist metamorphism during the early Devonian (Moseley, 1972 and this volume).

Wales

The Welsh Basin is now seen as a back-arc extensional basin within continental crust (cf. Dewey, 1969 and Figure 1.2). Its original relationship to the Lake District (and the Irish equivalents) is not clear. Between the two, lies the small, isolated area of the Precambrian rocks of Anglesey. The boundary of the Anglesey terrane with the Lake District terrane is not exposed. Its south-eastern boundary, with the Welsh terrane, which has previously been interpreted as a subduction zone active in early Palaeozoic times, is now seen (Gibbons, 1987) to be a fault boundary marking Late Precambrian strike-slip docking of the small terranes that now make up Anglesey. The significant deformation related to folds and cleavage in Wales represents, as in the Lake District, essentially an early Devonian event. There are, however, many variations on a simple pattern, attributed, variously, to soft-sediment, tectonic, and volcanic activity.

Two particular structural features have provoked discussion. Firstly, there is the arcuate pattern of folds and cleavage from E–W turning to N–S, which has been most commonly attributed to basement control, and regarded by Soper *et al.* (1987) as part of the same curvature as that seen between the Lake District and the Craven Inliers. Secondly, there is the diminution of deformation south-eastwards, seen as the diminished affects of the Caledonian Orogeny towards the south-east margin of the orogen, represented by the basement rocks of the Welsh Borders and the English Midlands.

CONCLUSIONS

Although there are uncertainties concerning the timing and mechanisms of the Southern Upland structures, the three areas are apparently united by their common history of cleavage formation and maximum shortening which climaxed in the early Devonian. Evidence is accumulating (Soper *et al.*, 1987; McKerrow, 1988; Soper, 1988) that this event may be equivalent to the Acadian Orogeny (Emsian in age) of the Canadian Appalachians. One feature of this cleavage that unites the two sides of the Iapetus suture, running through the Solway Firth, is the transection (cross-cutting) of folds by the cleavage, which has now been recognized widely in the Southern Uplands, Lake District and Wales. The Southern Uplands and the Lake District are also united by their flat-lying D_2 folds and cleavage, which may be related, in time and space, to the major granite intrusives that characterize both areas.

Another feature which unites the latest Caledonian deformation across the whole of Britain is faulting, much of which is strike-slip and much of that sinistral (Hutton, 1987). The faults range from the Great Glen Fault system in the Scottish Highlands to the Welsh Borderland Fault system (Woodcock and Gibbons, 1988). The minor faults, in the Scottish Highlands, the Southern Uplands, and the Lake District especially, commonly show a more NNE–SSW trend and sinistral displacement. These two features, cleavage transection and faulting, have been used by both Hutton (1987)

and Soper *et al.* (1987) to reconstruct the positions of the British Caledonian terranes and the relative movements and geometries of the margins of Iapetus itself.

Throughout the British Caledonides it is being increasingly recognized that certain structures, both folds and faults, have origins related to basin development that pre-date the main Caledonian structures. This recognition has not only allowed a clearer understanding of the early development of the area, but also removed some apparent tectonic ambiguities. For example, recent studies in both the Southern Uplands and the Lake District have recognized that certain folds are of soft-sediment origin. Similar folds have long been recognized in the Silurian of Wales (Woodcock, 1976) and other anomalous structures there (in older rocks) are also being attributed to this origin. Again, in Wales, early faults have been related to volcanic activity, as well as to facies and thickness changes. Comparable features are now being recognized in the Lake District.

In the Southern Uplands, the major strike faults are seen to have an early history that controlled the development of sedimentation in the accretionary prism, and recently, smaller-scale fractures have been attributed to shortening and extension in the accreting sediments.

Further details and references may be found in the 'Introductions' to the following chapters, which deal individually with the sites in the Southern Uplands, Lake District, and Wales.

Chapter 2

Southern Uplands

INTRODUCTION – A STRUCTURAL PERSPECTIVE
J. E. Treagus

The stratigraphical framework

For almost a hundred years, after the first researches of Lapworth (1874, 1889) and the work of the Geological Survey of Scotland (Peach and Horne, 1899), there was a curious lack of interest in the structure of the Southern Uplands. Today, the area is celebrated internationally for its structures, which are perhaps the most quoted examples of folds, cleavage, and faults developed in an ancient accretionary prism. In the early Palaeozoic, that prism lay just to the north of the presumed Iapetus suture.

The gross stratigraphical framework of the rocks is largely that established by Peach and Horne (1899), and is shown in Figure 2.1. The rocks are dominantly greywacke, with a smaller proportion of mudstone and shale, and a minor, but significant, amount of basic igneous rock associated with black shale. Although only sparsely fossiliferous, the rocks have been shown to range in age from the Arenig Series in the north to the Wenlock in the south. They generally exhibit a steep dip to the north-west or south-east.

The early workers clearly recognized the intensity of both folding and cleavage, although little use was made of minor structural geometry to define the major structures. Hall (1815), one of the first to realize the significance of folded strata in terms of the stresses imposed, was principally inspired by observations on the rocks of the Berwickshire coast. He carried out some of the first model experiments in an attempt to reproduce these folds.

The first structural cross-sections of the Southern Uplands (Lapworth, 1889; Peach and Horne, 1899) depicted the Ordovician rocks of the north as essentially the core of a broad anticline (Figure 2.2), south-east limb of which was corrugated by tight asymmetrical folds. These were seen principally in the Silurian rocks which were the subject of most of the following site descriptions. The folded sequence on the coastal section of Stewartry between Knockbrex and Kirkandrews, and described here in the Barlocco site, was particularly remarked upon by Peach and Horne (1899, p. 215). At the same section, they appear to have been aware that cleavage had a trend clockwise relative to that of the folds (Peach and Horne, 1899, p. 214), a relationship which has proved

critical to the plate-tectonic interpretation of the area. The sites are described in terms of the deformation phase history used by Stringer and Treagus (1980, 1981) and followed by most subsequent workers. The D_1 phase produced the upright ENE-trending folds which dominate the structure of the area. The D_1 folding may be diachronous, but the youngest rocks affected are the *C. lundgreni* biozone of the Wenlock. The S_1 cleavage is treated here as coeval with the D_1 folding but might entirely post-date it. The D_2 deformation, which is locally associated with folds that have flat-lying axial-surfaces and cleavage, post-dates dykes which have been dated from 403 to 392 Ma.

Structural observations

The 1960s saw the first important stage of structural reassessment. An examination of the sedimentary structures in the greywackes led Craig and Walton (1959) to realize that minor and major fold vergences were to the south-east, which, combined with the NW-younging of the long limbs of the folds, meant that the previous structural interpretation had to be erroneous. However, an interpretation was required that would reconcile these observations with the irrefutable fact that older rocks were encountered in successive belts from south-east to north-west. Walton (1961; see Figure 2.2) proposed a structure for the Southern Uplands which consisted of a series of north-facing monoclines, the northward descent of which was counteracted by a number of steep reverse faults with a southerly downthrow. These thrusts, often marked by the occurrence of black shale (the Moffat Shales) and interpreted as such by Peach and Horne (1899), essentially bound strike-parallel blocks which are of increasing age north-westward.

Further work in the 1960s by Rust (1965), Walton (1965), and Weir (1968) was concerned with more detailed description of the minor structures in the Silurian rocks of the Central Belt (Figure 2.1). This work proposed a number of deformation phases which were related to fold attitude and cross-cutting cleavages, but all emphasized the dominance of the asymmetrical, usually SE-vergent, open to tight folds with wavelengths of tens or hundreds of metres, which are now attributed to the first deformation phase (D_1). Toghill (1970) was responsible for showing the relationship between Walton's structural model and the constraining graptolite biostratigraphy.

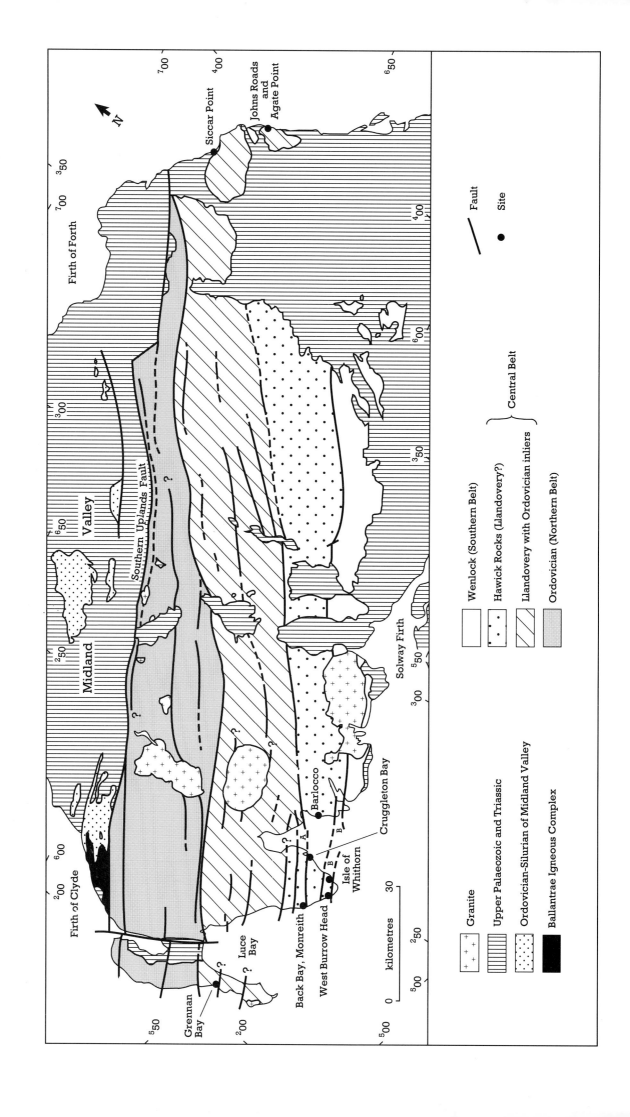

N

Firth of Forth

Siccar Point

Johns Roads and Agate Point

Midland Valley

Southern Uplands Fault

Firth of Clyde

Grennan Bay

Luce Bay

Back Bay, Monreith

West Burrow Head

Isle of Whithorn

Barlocco

Cruggleton Bay

Solway Firth

Fault

Site

Wenlock (Southern Belt)

Hawick Rocks (Llandovery?)

Llandovery with Ordovician inliers

Ordovician (Northern Belt)

Central Belt

Granite

Upper Palaeozoic and Triassic

Ordovician-Silurian of Midland Valley

Ballantrae Igneous Complex

kilometres

0   50   30

The only description of detailed structure in the Ordovician of the Northern Belt is that of Kelling (1961). Although some of the more elaborate structural histories have now been simplified (Weir, 1979; Stringer and Treagus, 1980, 1981), this work recognized two important structural features: firstly, the presence of steeply plunging and curvilinear hinges and, secondly, the presence of a superimposed local deformation, producing cleavage and axial surfaces with flat to moderate dips to the south-east.

Stringer and Treagus (1980, 1981) interpreted the deformation history of the Southern Uplands in terms, essentially, of a single deformation phase (D_1) with local modification by D_2. From observations principally in the Hawick Rocks, in the Wigtown Bay area, they described D_1 folds as generally tight and upright with first-order folds having amplitudes of about 300–500 m and spacing of 0.25–3.0 km. The folds of this scale are asymmetrical with vergence to the south-east, but intermediate folds, seen on a scale of one hundred metres or less, locally verge to the north-west in narrow belts on the short limbs of the major folds (see Figure 2.6A). Plunge is typically gentle to moderate (0–45°), both north-east and south-west. A belt of rapid plunge variation (50–90°), giving rise to local downward-facing, was identified near the southern margin of the Hawick Rocks.

A major feature of interest to structural geologists was the observation by Stringer and Treagus (1981) that the cleavage (S_1) related to the first deformation was not parallel to the axial surfaces of individual folds. This feature is now widely recognized elsewhere in the Caledonides, and is referred to as 'fold transection by cleavage'. It is manifested, in the Southern Uplands, by an approximate 10° difference in the strike of cleavage, clockwise with respect to that of the axial surfaces. Parallelism, as well as occasional anti-clockwise transection, was also observed. Stringer and Treagus demonstrated that, although the customary divergent (in mudstone) and convergent (in sandstone) cleavage fans were developed

around the folds, the transection resulted in inter-section lineations, between bedding and cleavage, in both materials which were dispersed, as shown in Figure 2.6B, from parallelism with fold hinges.

The D_2 deformation, as defined by Stringer and Treagus (1980), is exposed as open to tight folds with wavelengths from tens of centimetres to tens of metres, associated with a flat, south-easterly dipping, crenulation cleavage. These folds affect steep-dipping D_1 fold limbs resulting in generally subhorizontal plunge and neutral vergence. These folds are strongly developed in a 2–3 km-wide belt in the northern part of the Hawick Rocks across the Wigtown Peninsula.

The accretionary prism model

Dewey's (1969) paper on the evolution of the Caledonides (Figure 1.2) awakened interest in the Southern Uplands area for its proposed position on the continental margin above the northward subducting plate that was the floor of the Iapetus Ocean, but it was not until the paper by Mitchell and McKerrow (1975), comparing its evolution with that of recent accretionary prisms that workers began to look for more detailed structural analogies. Mitchell and McKerrow (1975), McKerrow *et al.* (1977), Leggett *et al.* (1979), and Leggett (1980) particularly emphasized the similarities of the stratigraphical arrangement of the Southern Uplands to modern examples. The accretionary prism described from many examples of modern destructive plate margins, consists of a sequence of sedimentary units separated by thrusts dipping towards the continent; each sedimentary unit youngs upwards, but the age of successive units within the thrust 'prism' increases towards the continent. This arrangement, with belts of rocks of increasing stratigraphical age to the north-west, within which rocks show consistent younging towards the north-west, had essentially been established by Walton (1961). He had also shown that many of the belts were bounded by steep faults, downthrowing to the south, and generally referred to as thrusts. The role of the thrusts that separate the belts is particularly important in the accretionary model (Figure 2.2), as is the distinct chronostratigraphical make-up of each thrust-belt. The geometry of the D_1 folds, within the thrust slices, was also emphasized by Stringer and Treagus (1980, 1981), the original flat south-easterly (oceanward) asymmetry being regarded as a result of deformation above the north-

Figure 2.1 Geological map of the Southern Uplands, showing the distribution of the three main belts, some of the steep faults that bound these belts, and subsidiary tracts. The positions of the sites discussed are also shown. A and B, in the south-west, show the zones of D_2 folding and steep D_1 plunge respectively, as discussed in the text.

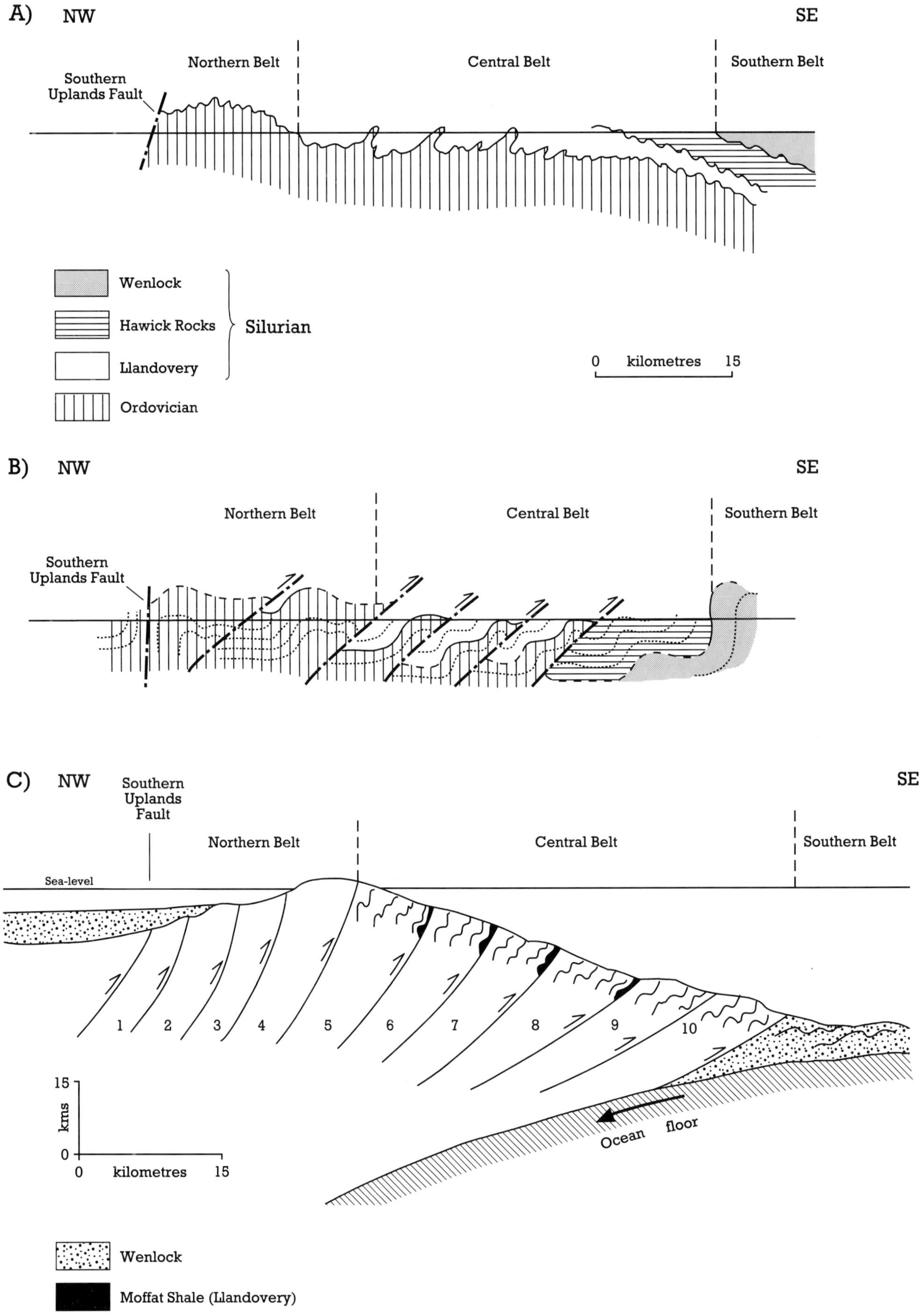

A)

Southern
Uplands Fault

Northern Belt | Central Belt | Southern Belt

	Wenlock	
	Hawick Rocks	} Silurian
	Llandovery	
	Ordovician	

0   kilometres   15

B) NW   SE

Northern Belt | Central Belt | Southern Belt

Southern
Uplands Fault

C) NW   Southern
Uplands
Fault   SE

Northern Belt | Central Belt | Southern Belt

Sea-level

1   2   3   4   5   6   7   8   9   10

Ocean floor

15
kms
0

0   kilometres   15

	Wenlock
	Moffat Shale (Llandovery)

westerly subducting ocean-floor with subsequent rotation, to a steep attitude, within the thrust-bound packets. The plunge variations of the folds was also attributed to proximity to the thrust zones.

Detailed structural observations on the thrusts themselves have been sparse, partly owing to their poor exposure. Until recently (for example, Rust, 1965; Toghill, 1970; Fyfe and Weir, 1976; Cook and Weir, 1979), work has concentrated on thrust geometry and stratigraphical separation, and observations on shear-sense or timing, with respect to D_1, have been rare. Webb (1983) considered that thrusts in the Ettrick imbricate structure resulted from south-easterly-directed extension of rotated short limbs of the D_1 folds, but that they pre-dated the S_1 cleavage.

Stringer and Treagus (1981) and Treagus and Treagus (1981) have attempted to relate the non-axial plane cleavage in the Southern Uplands to this plate tectonic setting. They pointed out that a non-orthogonal relationship between the initial folds in the sedimentary pile and the subsequent stresses, caused by oblique ocean closure, would result in the observed transection of folds by cleavage. Other workers (for instance, Sanderson *et al.*, 1980) have interpreted the phenomena in terms of transpression, essentially a combination of pure shear perpendicular to the sedimentary strike and simple shear (sinistral) parallel to that strike, as a consequence of oblique convergence of the margins of the Iapetus. Such a model predicts horizontal stretching near the ocean suture and down-dip stretching away from the suture. Such strain variations are recorded in the Irish Caledonides, but in the Scottish Southern Uplands, Stringer and Treagus (1981) record only very weak down-dip extension in the flattening strain associated with cleavage. Thus the obliquity between cleavage and the axial planes of folds is particularly important evidence for the model of the Southern Uplands as a part of an ancient accretionary prism. More precise strain measurement will help in the clarification of the model.

Figure 2.2 Cross-sections of the Southern Uplands. (A) After Lapworth (1889); (B) after Walton (1961); (C) reconstructed profile of the accretionary prism, in Wenlock times. The tracts 1–10 are of decreasing age south-eastward, within each tract rocks young to the north-west. The style of the D_1 folding is shown schematically in the Llandovery and Hawick Rocks of the Central Belt (after Leggett *et al.*, 1979).

Several workers have identified isoclinal folds which appear to be pre-D_1 folding, particularly in the Hawick Rocks. Rust (1965) was the first to appreciate the presence of such large-scale isoclinal folds and their presence has been confirmed in several of the present site investigations. Such folds, as well as disrupted units and other slump structures, are identified by Knipe and Needham (1986) as an essential part of the early evolution of accretionary complexes.

Recent work

There has been a spate of work in the Southern Uplands since 1986, much of it stratigraphical and sedimentological, which has resulted in variations on and repudiation of the accretionary prism model. Some papers (see below) provide important new structural data, which undoubtedly will be used in future site selection. Barnes *et al.* (1987), for example, report new work in the Rhinns of Galloway (McCurry) and on the Wigtown Peninsula (Barnes), in previously undescribed rocks of the Central Belt. Unusual aspects of this work are the description of a substantial belt of north-verging D_1 folds in SE-younging rocks, with northward downthrow on steep faults on the Rhinns. Thrusting, both north and south down-throwing, is syn-D_1, as are the steeply plunging folds, except where rotated locally by post-D_2 deformation. Kemp and White (1985) and Kemp (1987) have detailed, for the first time, some Southern Belt rocks of Wenlock age. Apart from new palaeontological and sedimentological data, which reveal the highly imbricated nature of the northern part of this belt, he discusses the nature of the 'sheared zones' which characterize the rocks: the bedding is imbricated and disrupted, and they are affected by SE-verging folds, faulting, and boudinage. The origin of these zones appears to be partly soft-sediment deformation, but essentially they formed during D_1, partly post-dating S_1. Common, sinistral, steep-plunging folds post-date D_1.

The attack on the interpretation of the Southern Uplands as a forearc accretionary prism especially by Hutton and Murphy (1987), Morris (1987), and Stone *et al.* (1987) is mentioned above. The detailed scrutiny of greywacke provenance and palaeocurrents has shown that there was a significant Ordovician input of sediment from the south, much of it derived from a missing volcanic arc, that is an arc destroyed through subduction or strike-slip faulting (see below). These and other, essentially sedimentological, arguments are certainly going to produce a reassessment of the structural evolution of the Southern Uplands in future years.

The variations in the accretionary prism model, for instance the role of imbricate thrusting and the timing of collision (Hutton and Murphy, 1987; Stone *et al.*, 1987), may eventually be resolved by the detailed examination of the structures such as the geometry of the D_1 folds, the timing of the transecting S_1 cleavage, the strain variation, particularly towards and within the thrust zones, the significance of the plunge variation, and especially the relationship between thrusting and faulting and other deformation events. Of particular interest will be a comparison of the structural history and style of the Northern Belt with that of the Silurian rocks. Hutton and Murphy (1987) report cleaved Middle Ordovician (*gracilis* biozone) shales as clasts in the Silurian, and Morris (1987) describes an earlier deformation history in the Ordovician of the Longford Down Inlier in Ireland.

As yet, the most significant structural contribution to the new models has come from Anderson and Oliver (1986), in a reinterpretation of the displacement on the Irish Orlock Bridge Fault. The apparent equivalent of this fault in Scotland, the Kingledores Fault, separates the Ordovician of the Northern Belt of the Southern Uplands from the essentially Silurian rocks of the Central Belt (Figure 2.1). It features in the accretionary prism interpretations (for example, Leggett *et al.*, 1979) as one of the major thrust boundaries. From evidence, principally in the Irish outcrops, Anderson and Oliver (1986) demonstrate that, whatever its early history, the fault suffered significant sinistral strike-slip movements in the Late Silurian and they argue that these movements were of the order of 400 km. Hutton and Murphy (1987) claim that this fault was responsible for the removal of the missing volcanic arc, mentioned above, and Hutton (1987) argues that it is one of the several major, strike-slip terrane boundaries which have featured in the evolution of the British Caledonides.

Another of these terrane boundaries would be the Southern Uplands Fault, the northern boundary of the rocks treated in this volume. Traditionally, it is seen as a steep fracture (with major splays into the Glen App, Stinchar Valley, and Lammermuir Faults). It is assumed to have a downthrow to the north-west, based on the contrast of Lower Palaeozoic facies on its two sides and on the truncation of the Lower Devonian on its northern

side. In the accretionary prism models, the fault is usually shown as a successor to one of the powerful, steep, NW-downthrowing thrusts, but Bluck (1986) also suggests that it is the site of NW-thrusting which has caused the juxtaposition of the trench sediments of the Southern Uplands against the proximal forearc deposits of the Girvan area to the north. Again, from consideration of sources of Lower Palaeozoic sediments north and south of the fault, Hutton (1987) argues for substantial strike-slip motion in the Late Silurian. Exposures of the fault zone are highly brecciated, fractured and veined, but there are no reports of any local criteria which can be used to prove the Caledonian displacement sense.

Faults

Minor faults of post-D_1 age, which are probably Caledonian in age, abound in the Southern Uplands. Many workers report brecciation and other symptoms of brittle deformation coincident with the major strike-parallel thrust faults, as well as those at high angles to the strike. The N–S wrench faults, with sinistral displacement, are particularly widely reported. Dextral wrench faults with a more north-east trend are also reported (Weir, 1979). Other post-cleavage faults and fractures, showing a wide range in orientation, can be seen in most of the sites described; many clearly have a displacement, but the absence of markers makes a unique calculation of displacement sense, or amount, difficult. Low-angle faults, with both north-west and south-east dips, with both thrust and normal displacement, and throws apparently in the order of a few centimetres, can be deciphered in most shore sections. No overall stress vector pattern has been proposed.

Timing of deformation

The precise timing of the various stages of the Caledonian deformation in the Southern Uplands is not always clear. In particular, there are dates for neither the development of the D_1 folding, which might be expected to be diachronous across the fold belt, nor for the cleavage development. However, D_1 folds and S_1 cleavage affect rocks of at least the *C. lundgreni* biozone of the Wenlock Series. Dykes which post-date S_1 and largely pre-date D_2 folds (Stringer and Treagus, 1980) have been dated (Rock *et al.*, 1986) between

418 and 395 Ma. These dykes are mostly cut by the Devonian granites which have dates ranging from 408 to 392 Ma. The writer's observations on contact porphyroblasts suggest that the Cairnsmore of Fleet Granite (392 ± 2 Ma) pre-dates, or is closely associated with, D_2: this relationship is very similar to the relationship seen between the granites and folds (D_2) in the Lake District. Lower Devonian lavas (Upper Gedinnian) unconformably overlie folded Silurian at St Abb's Head as well as in the Cheviots, where the lavas are cut by the Cheviot Granite dated at 391 Ma. Although these lavas are often quoted (Powell and Phillips, 1985) as Upper Gedinnian in age, McKerrow (1988) points out that both the faunal evidence and the latest radiometric dates (389–383 Ma; Thirlwall, 1988) would allow a Late Emsian age (perhaps about 397–390 Ma) for the main deformation event. Such a date would correlate well with that deduced for the Lake District and Wales (Soper *et al.*, 1987; Soper, 1988; McKerrow, 1988) and with part of the Acadian Orogeny of Canada.

The selected sites are shown in Figure 2.1. In spite of the detailed sections of the Northern Belt by Kelling (1961), it was not possible to select any site in these rocks which would illustrate their typical D_1 style. It is hoped that future work will especially clarify the relationships of the cleavage and thrusts to the folds. All the sites selected thus far are located in the Central Belt greywackes and mudstones of Llandovery (presumed) and Wenlock age, and all exhibit the dominant NE–SW strike and steep dip that characterizes most of the Southern Uplands.

All the sites (except for Burrow Head) also show the characteristic NW-younging on the long limbs of the SE-verging D_1 folds, as well as transection of the folds by the S_1 cleavage. One site (Back Bay, Monreith) illustrates the style of the deformation, and two (Burrow Head and Grennan Bay) illustrate the nature of the faulted junctions. No suitable site was found to illustrate the later faulting, including the Southern Uplands Fault and parallel fractures.

SICCAR POINT
(NT 81187100–81307095)
J. E. Treagus

Highlights

Siccar Point is one of the world-renowned localities where Hutton (1795) first recognized the significance of unconformities in the geological record. In the context of this volume, it exemplifies the style of folding in the Silurian, which led Hall (1815) to make his important deductions concerning the relationship of stress to the formation of folds, through experiments.

Introduction

This coastal site exposes the angular unconformity between: 1. beds of Llandovery greywacke and shale; 2. beds of Upper Old Red Sandstone (ORS) breccia and sandstone (Figure 2.3). The tight folding and cleavage of the end-Caledonian deformation, seen in the Silurian, contrast strikingly with the gentle dips of the ORS. The breccias contain fragments of cleaved Llandovery rocks. The site was one of the localities at which the significance of unconformities in the geological record was first appreciated (Hutton, 1795; Playfair, 1805). Moreover, the folds in the Silurian of this coast, so well seen in the site, are those which inspired Sir James Hall (1815) to undertake his early experiments in model rock deformation.

Description

The Silurian strata, exposed over about a $100 \, \text{m}^2$, are folded in a tight synform, whose limbs dip steeply south, with an interlimb angle of $25°$. The fold plunges $35/240°$, with its south-east limb overturned and an axial-surface attitude of $080/66°\text{S}$. Cleavage in siltstones and shales, inter-bedded with the greywackes, has an attitude $105/74°\text{N}$ and thus transects the axial surface. Way-up structures (bottom structures, graded bedding, and ripple cross-lamination) show the fold to be an upward-facing syncline.

The Old Red Sandstone beds dip gently north; the unconformity surface, although broadly parallel to this dip, exhibits local irregularities due to differential erosion (often along strike) of the underlying greywackes and shales (Figure 2.3). These irregularities cause certain greywacke beds to protrude several metres above the principal planar unconformity surface. Indeed, the whole synclinal fold, described above, is one such major irregularity protruding above the unconformity surface.

Figure 2.3 Siccar Point. Subvertical Silurian greywackes and cleaved shales on the south limb of a tight Caledonian syncline are unconformably overlain by Upper ORS breccias and sandstones. View looking east with lens cap (centre) for scale. (Photo: J. Roberts.)

Interpretation

The principal interest of this site is that it displays, in a unique three-dimensional manner, Silurian rocks, folded and cleaved during the Caledonian Orogeny, eroded probably during mid-Devonian times and subsequently overlain by the terrestrial Upper Old Red Sandstone. However, it also exemplifies the style of D_1 deformation seen locally. There is no modern description of the structure of this coast, except where the folds show the complex curving hinges (Dearman *et al.*, 1962), described at John's Road also in the Central Belt. It is, therefore, of interest that this site shows very similar features to those described to the south-west, showing the persistence of such features, especially the same clockwise transection of folds by cleavage.

This site is the only locality in the Southern Uplands where the three-dimensional nature of the unconformity can be clearly seen. The contrast between the vertical greywacke and the horizontal red sandstones led to an understanding, not only of the fundamental earth movements that rocks undergo (Hutton, 1795), but also to an appreciation of the scale of erosion involved and thus to the immensity of geological time. The style of the folding along this coast, exemplified by this site, also led Sir James Hall (1815) to perform his early experiments and to make important deductions concerning the relationship between stress and folding, specifically that horizontal crustal shortening could be responsible for folding.

The unconformity at Siccar Point, though renowned for Hutton's appreciation of the significance of the phenomenon, does not in fact constrain the 'end-Caledonian' climax in the Southern Uplands very tightly. Elsewhere, in the south-western Southern Uplands, Wenlock rocks are unconformably overlain by Upper ORS sediments and intruded by post- or syn-deformation granites dated about 400 Ma. Nearer the present site, at St Abb's Head, folded Llandovery rocks are unconformably overlain by Lower Devonian lavas, and in the Cheviots probable Wenlock rocks are overlain by lavas of the same age. The possible age of the Caledonian deformation events is discussed in the Introduction to this chapter.

Conclusions

This site has been included as a unique and historic locality in the Southern Uplands to demonstrate the unconformity between the Silurian greywackes, strongly deformed in the Caledonian Orogeny, and the flat-lying undeformed Upper Devonian red sandstones. It also illustrates the D_1 deformation style of the north-eastern exposures of the Central Belt; the truncation of D_1 folds, at the locality, by the flat-lying Upper Old Red Sandstone breccias and sandstones is the clearest example in the Southern Uplands, of the timing of the Late Caledonian deformation. It allows a comparison with the similar style of deformation seen in exposures to the south-west at Barlocco, Cruggleton Bay, and West Burrow Head; that is, the asymmetry of the fold, the attitude of its limbs, axial-surface, fold hinge, and the clockwise transection of the cleavage.

JOHN'S ROAD AND AGATE POINT (NT 95286411–95476410)

J. E. Treagus

Highlights

The plunge variations at John's Road are extraordinary, by the standards of any slate-belt folding, and are certainly the best exposed and most dramatic yet described in the 'non-metamorphic' Caledonides. The plunge variations occur in a zone some 300 m wide, but the regional context is not known.

Introduction

This site is located in the Llandovery rocks of the Central Belt (Figure 2.1) in the poorly described Berwickshire coastal section. Originally reported in the Geological Survey Sheet Memoir (Geikie, 1863), they have subsequently attracted the detailed description by Dearman *et al.* (1962). The regional setting of this coast section appears to be broadly similar to that of the Hawick Rocks in the Scottish south-west coastal sections (see below).

Unfossiliferous Silurian greywacke siltstone and cleaved shale generally dip steeply to the north-west and young in the same direction, but in detail are affected by folds, of a variety of scales, with gentle plunge and south-east vergence.

Dearman *et al.* (1962) described a zone on this coast, some 250 m wide across strike, in which the regional fold pattern alters to one in which fold plunge changes not only to vertical and steeply downward-facing (as in the Isle of Whithorn), but also to gently plunging, downward-facing attitudes. Agate Point and John's Road are the two principal localities which Dearman *et al.* (1962) described within this zone.

Description

In the 4 km coastal exposures, between Eyemouth and Burnmouth, the rocks show the typical structural features of this belt of the Southern Uplands, with steep or north-west-dipping and north-west-younging sequences, interrupted by folds with wavelengths of between 5 m and 20 m. These folds are open to tight, south-east-verging, with gentle plunges to the south-west or north-east, and are upward-facing on subvertical axial surfaces. In the Eyemouth area, immediately north-west of the site, fold plunges of 20–40°SW are characteristic, with variations up to 50°SW and 20°NE. About 100 m north of John's Road, plunge values to the south-west begin to increase; no obvious planes of movement are associated with this change. The plunge variations continue in a zone of some 300 m with a return to more usual plunges to the south of Agate Point. The zone may, in fact, extend beyond this point, but no data are available in the rather inaccessible cliff section to Burnmouth. Two localities have been selected in this zone, described in detail by Dearman *et al.* (1962), because they provide the best exposures of the full plunge variation.

Agate Point (NT 95416411)

This area, some 25 m by 50 m, illustrates the typical plunge behaviour of the folds; the folds (Figure 2.4) can be followed on to the islands to the north-east of the site. The map is based on Figure 3 in Dearman *et al.* (1962). Four greywacke units were mapped around the axial surfaces of four folds (1–4 on map). The folds are tight, with limb dips about the vertical, striking NNE or NE. Sedimentary structures provide clear evidence of the direction of younging, as indicated.

Plunge varies within the site, from 20° upward facing, to 28° downward facing. Fold 1 shows the most marked and rapid variation from 70°NE (downward-facing), through the vertical to 50°SW (upward-facing) over some 15 m. To the north-

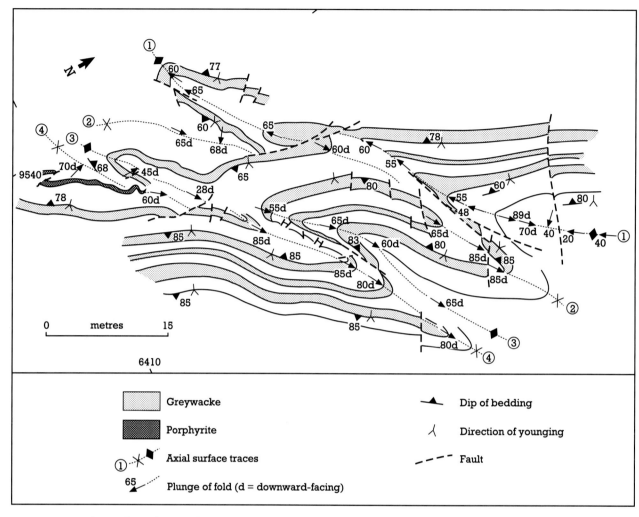

Figure 2.4   Geological map of Agate Point (after Dearman *et al.*, 1962).

east, this fold can be traced to beds where hinges plunge as low as 20°SW. Folds 2, 3 and 4 all show similar variations, from vertical to 45°NE (downward-facing). Discontinuities occur on the fold limbs (see map) and may well have been essential, as zones of high shear strain, to the mechanism whereby the steep plunges were achieved.

John's Road (NT 95366414)

This locality is one rib of rock, some 40 m by 10 m, in a broader zone of plunge variation. The exposure illustrates, particularly well, plunge variation in three dimensions. The exposure is well documented in Dearman *et al.* (1962, Figure 5;

see Figure 2.5, this volume). The greywacke–shale sequence here is folded into a prominently displayed anticline–syncline pair, which plunges steeply south-west. Successive fold hinges, of this fold pair, can be followed to the north-east where they rapidly pass through the vertical and assume low plunges (30° and less) towards the north-east. These folds can be shown, from the younging evidence, to be downward-facing and provide a most dramatic illustration of the plunge variation. Again, discontinuities can be observed on fold limbs and these, together with beds of disrupted material, may provide evidence of the origin of the plunge variation.

In both localities, the shale beds are affected by a weakly developed cleavage, which is NE–SW-trending and broadly parallel to the axial surfaces of the folds. No mineral fabric, or other exceptional

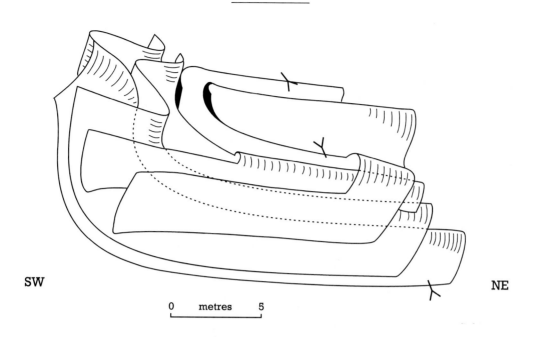

SW

NE

0   metres   5

Figure 2.5 Diagrammatic representation of the folds at John's Road (after Dearman *et al.*, 1962).

textural features of finite strain, have been observed in the field, which might be used to explain the exceptional fold plunge variation.

Interpretation

The interest of this site lies not only in the unusual range of plunge variation, but in the short distance over which it occurs. Within the Southern Uplands, the features exhibited in the site are unique; comparable examples have not been described. Somewhat similar features have been reported by Stringer and Treagus (1980, 1981) and the Isle of Whithorn site described in this volume; in these instances, however, the plunge variation is neither as extreme, nor is it seen to occur over such short distances.

Dearman *et al.* (1962) attribute the plunge variation to refolding, claiming that 'it can be demonstrated that folds, with a NE–SW axial trend, have been refolded about NW–SE axes and that the two show a common axial plane' (p. 275). In discussion of the paper by Dearman *et al.* (1962), Westoll (p. 283) comments, that such a phenomenon might be a consequence of 'a single continuous movement picture' resulting from the directions of expulsion of intergranular water. Other speakers comment on the lack of superimposed minor structures and the possible contribution of wet-sediment deformation (pp. 284–5).

Stringer and Treagus (1980 pp. 328–9), in their discussion of plunge variations seen in a somewhat similar zone in the south-west Southern Uplands, appeal to heterogeneous strain both within the zone and of the zone itself relative to the rocks outside, and to the physical rotation of packets of folds bounded by shear planes.

There has been widespread interest in reports of strongly curvilinear and 'sheath' or 'eyed' folds in recent years. These reports (Roberts and Sanderson, 1974; Cobbold and Quinquis, 1980) have been based on observations in the field of modelled folds where strong extensional strains are observed. However, there is no evidence of such strong extensional strains, or indeed, of any unusual strain pattern in the rocks where these phenomena are reported from the Southern Uplands.

The rocks of this site do exhibit two features which may be important in future research. Some sequences are noticeably disturbed and individual beds cannot be traced far along strike. This may be due to bedding-plane movements during folding and, certainly, there are also discordant zones along which there is similar disturbance. However, it is possible that the fold plunge variation might be partly attributed to the existence of soft-sediment folding (cf. Webb and Cooper, 1988, describing similar situations in the Lake District), which pre-dated the tectonic folding and with which the local chaotic bedding might be associated. Irregularity of bedding leads to discontinuous beds which furthers variation in plunge.

In the part it has to play in discussion of the accretionary prism model that has been advocated for the Southern Uplands (see Chapter 1), this site may be important in two respects. Firstly, this zone of unusual plunge variation, like that to the south-west, may be indicative of local strain gradients associated with thrusting (Stringer and Treagus, 1980) and secondly, it may be significant in the clear association cited by authors of wet-sediment movement within the accretionary prism setting (Kemp, 1987; Knipe and Needham, 1986).

Conclusions

This site is included, primarily, for its particularly clear exposures of local zones of unusual complex folding. These contortions were perhaps produced by the intense compressive stress that was generated during the Caledonian mountain-building episode (orogeny). In general, fold structures elsewhere have a fairly regular form, but at this locality the hinges of the folds are strongly curved, sometimes by as much as 165°. This is exceptional not only in the Southern Uplands but in the Caledonian Orogenic Belt as a whole. These zones are important in understanding the evolution of the large-scale structure of the Southern Uplands, as indicating the presence either of local strain gradients associated with thrusting or horizons which suffered wet-sediment deformation, the latter associated with the accretionary prism which, prior to the orogeny, existed on the northern side of the Iapetus Ocean.

BARLOCCO (NX 58054880–58904820)
J. E. Treagus

Highlights

The coastal section, which is easily accessible and continuously exposed, illustrates all the typical features of the Caledonian D_1 folding and cleavage in the Hawick Rocks, and the geometry of D_1 folds in the Central Belt of the Southern Uplands.

Introduction

This site is part of the shore exposures between Knockbrex and Kirkandrews (Figure 2.6A), which are cited by Peach and Horne (1899, p. 215) and Craig and Walton (1959). The sites were important in the formation of the views of these authors on Southern Uplands structure. Craig and Walton (1959) used this section to illustrate their theory that the Southern Uplands structure comprised a number of large monoclines with alternating steep, relatively unfolded, limbs and flat zones in which a small thickness of rocks was repeated in recurrent symmetrical folds. The West Burrow Head site lies in part of one of these steep limbs, but the Barlocco site is part of a flat zone. Stringer and Treagus (1980, 1981) have contested this view, claiming that south-east verging fold pairs are evenly distributed in the Hawick Rocks and generally indicate a moderate sheet-dip to the north-west. Craig and Walton (1983) have defended their view.

The typical features of the D_1 deformation in the Hawick Rocks, seen in this site, are very similar to those in other adjacent Llandovery and the Wenlock formations. These features have also been described in papers by Stringer and Treagus (1980, 1981), where this section is mentioned in particular. Most of the characteristics of Southern Upland folding, described by Walton (1983), can be observed in this section. The typical features are: the wavelength, asymmetry, and plunge variations of the folds and the transection of these folds by the S_1 cleavage.

Description

The principal feature illustrated in this site is one common to much of the Southern Uplands, that of steeply dipping greywackes, younging to the north-west, but periodically interrupted by small-scale fold pairs. In this section, alternating greywackes (0.3 m to 3 m thick) dip steeply to the north-west or south-east. Sedimentary structures – grading, loading, and cross-stratification – are usually found to demonstrate the dominant north-west younging. About twenty major fold pairs are exposed, with an average wavelength of 42 m. Minor folds with wavelengths down to 5 m also occur, but shorter wavelengths are rare.

The folds are mostly tight to moderately tight (interlimb angle 20–70°) with axial surfaces which are subvertical. Hinge style varies according to lithology, thickness, and position in a fold stack; some in massive greywackes are concentric, but more typically alternating greywacke, siltstone, and

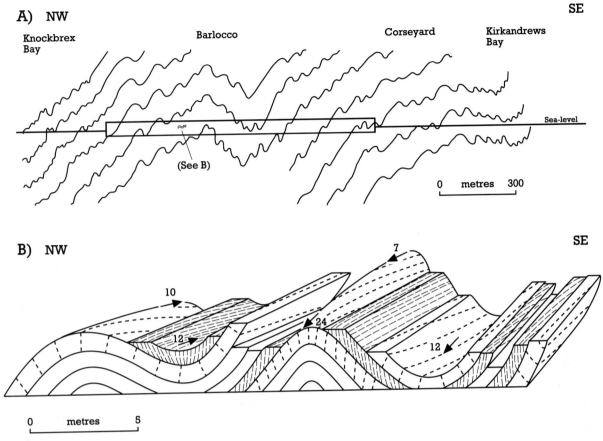

Figure 2.6 (A) Diagrammatic fold profile of the Knockbrex Bay–Kirkandrews Bay coast section, with box indicating the location of the Barlocco site. Approximate position of the folds illustrated in (B) is also shown. (B) Typical fold and cleavage geometry at the Barlocco site, based on field observations at NX 5835 4865. Cleavage is shown: open spaced in sandstones and narrow spaced in mudstone. Plunge of fold hinges and cleavage–bedding intersections are also shown (after Stringer and Treagus, 1980, figure 2).

mudstone multilayers show a style intermediate between chevron and 'similar' (that is, Class 1C, Ramsay, 1967).

The ratio of the width (measured horizontally) of the north-west-younging limbs to the south-east-younging limbs of these folds is, on average, 4:1, a reflection of the asymmetry of these south-east verging fold pairs (Figure 2.6A). It is estimated that, if it could be assumed that faulting was unimportant in the section, the sheet dip is about 45° to the north-west.

It is clear, however, that the fold section is interrupted by a great number of fractures, some subvertical strike-parallel or dip-parallel, others dipping at low angles. Occasionally, it is possible to demonstrate apparent displacements of a few centimetres to a metre, but the absence of marker horizons makes matching across possible faults difficult and calculation of the precise movement sense problematical. One steep dip fault is

particularly obvious in the section, from the displacement of the folds on the upper foreshore relative to those on the lower foreshore.

Another feature of the site is the variation in plunge of the D_1 hinges, and the consequent variation in fold profile and the impersistence of individual hinges. Many hinges are sub-horizontal, but others plunge up to 30°, both to the north-east and south-west, and occasionally, across the foreshore, individual hinges can be seen to display this range within strike lengths of 10 m or more.

This site can also be used to examine the essential features of the S_1 cleavage in the Hawick Rocks. In sandstones, the cleavage is formed by parallel or anastomosing partings of 0.01–0.05 m. A weak, preferred, dimensional orientation of the quartz or feldspar is sometimes apparent, and dark seams, mica and opaque mineral concentrations, are the result of pressure solution. Even in the cleaved mudstones, the cleavage is domainal, with closely-spaced (0.1 mm and less) dark seams with strong mica orientation which separates paler microlithons with less-strongly oriented mica. In graded beds, cleavage shows gradations in intensity as well as refraction in dip. In profile view (Figure 2.6B), the S_1 cleavage forms strongly convergent fans in folds of sandstone, centred on the axial surfaces. The cleavage in mudstone sometimes displays a finite neutral point in a hinge zone (Ramsay, 1967, p. 417).

In this site, one of the most interesting features is the transection of the folds, in plan view, by the S_1 cleavage. The section allows many individual hinges to be examined in three dimensions and it is revealed that the strike of the cleavage is clockwise to that of the axial surfaces. This angle, between the subvertical axial surfaces and sub-vertical cleavage in the mudstones, is commonly about 10°, but in all lithologies ranges up to 25°. Since, in profile view, the cleavage exhibits the conventional fanning and refracting relationships to the axial surfaces, the three-dimensional relationship of both individual cleavage planes and their intersection with bedding, as they cross fold hinges, must be quite complex – see Figure 2.6B and Stringer and Treagus 1980, for details. The principal difference from conventional cleavage and bedding relationships is that their intersection can be seen curving clockwise across fold hinges to become steeper on the limbs. Great care has to be taken in using the relationship between cleavage and bedding (in plan view) in the conventional way to interpret fold geometry.

Interpretation

The coastal section at Barlocco has been drawn upon by several workers in their interpretation of the Southern Uplands. In Peach and Horne (1899, p. 215) it is one of the few coastal sections remarked upon and then for the intensity of the folding exhibited. Craig and Walton (1959, 1983) and Walton (1961) used this coast as a model for their influential theory that the structure of the Southern Uplands comprises large-scale mono-clines descending towards the north-west, made up of alternating steep belts of relatively unfolded rocks and flat belts (as at Barlocco) of strongly folded rocks.

Stringer and Treagus (1980, 1981) also refer extensively to this coastal section, but were unable to substantiate the monocline model (Stringer and Treagus, 1983). Instead, they maintain that the larger scale of D_1 folding in the Hawick Rocks is of 0.25–3.0 km wavelength, with long, north-westerly dipping limbs with a sheet dip of about 45° alternating with short limbs dipping steeply south-east (Figure 2.6B). Smaller-scale folds corrugate the limbs of these structures, seen at Barlocco. The asymmetry of the larger folds and the smaller-scale folds on the long limb is typically to the south-east.

Rust (1965) and Weir (1968) have also described the folding and cleavage of the Hawick Rocks, in adjacent coastal sections. They produced more complex histories for the development of the deformation, described as D_1 by Stringer and Treagus (1980, 1981), which would seem to result from a difference in interpretation of the cleavage relationship to the folds. In particular, these workers would relate the transecting cleavage to an entirely superimposed deformation (but see Weir, 1979).

The history of the development of the folds and cleavage, such as that described from this site, and its relationship to external mechanisms is of great current interest. McKerrow *et al.* (1977) and Leggett *et al.* (1979) view the stratigraphical arrangement of the Southern Uplands in the context of an accretionary prism model and note that the north-west-verging monoclines of Walton (1961) are opposite to those seen in modern accretionary prisms. Stringer and Treagus (1981) suggested that the style of folding observed by them (steep south-east vergence) was consistent with the ocean-verging, initially recumbent attitude expected in accretionary complexes, subsequently rotated into their present attitude in association with imbricate thrusting. Knipe and Needham (1986), working in an adjacent coast section, have identified disrupted rocks, thrusts and thrust-related folds which they show are similar to those described from modern accretionary prisms. Such features may well be present in the Barlocco site.

Conclusions

The Barlocco site has been selected as superbly exemplifying the principal features of the main Caledonian deformation (D_1) in the Southern Uplands. Rocks deposited in the seas of the early Silurian Period are here deformed by folds and cleavage planes (which in plan view are seen to cut across the folds). These features were produced by the intense compression of this area resulting from the convergence and final collision of the 'North American' and 'European' continents, which was responsible for the building of the Southern Uplands. This Caledonian mountain-building episode (orogeny) culminated around 400 million years before the present, at the end of the Silurian Period or early in the following Devonian. Although other locations might have been chosen, this site has the additional attractions of being historically important, easily accessible, and of being unaffected by subsequent phases of deformation.

CRUGGLETON BAY NORTH (NX 47704981–48504998)

J. E. Treagus

Highlights

Three folds on the foreshore of the north side of Cruggleton Bay illustrate one of the most interesting features of the D_1 deformation in the Hawick Rocks, that of folds transected by a contemporary cleavage. Cruggleton Bay also illustrates the typical geometry of D_1 folds in the Central Belt of the Southern Uplands.

Introduction

The rocks of the Cruggleton Bay area are the typical greywackes, siltstones, and mudstones of the Hawick Rocks. Their structure was first described by Rust (1965), although this coastal section must have provided inspiration for Lapworth's (1889, Figure 3) influential cross-section of the south-west Southern Uplands. In his discussion of the deformation of the Whithorn area generally, Rust (1965) clearly regarded folds, such as those seen at Cruggleton Bay, as D_1 (his F_1), but the cleavage that crosses them in a clockwise sense he considered to be later (his F_3). Stringer and Treagus (1980, 1981) interpreted the cleavage as essentially contemporary with the folds, although having a non-axial plane (transecting) relationship to them.

The site is not as continuously exposed as adjacent parts of the coast, but it offers the best opportunity to observe the full three-dimensional relationships between D_1 folds and the clockwise transecting cleavage (Figure 2.7).

Description

The exposures of interest are seen in three anticlinal folds which protrude from the low-lying foreshore (Figure 2.8A).

Fold A (NX 47704981; see Figure 2.8B) illustrates the geometry of the non-axial planar cleavage, developed in a slightly reddened mudstone in the hinge area of a fold. The fold is a tight anticline, plunging gently about the horizontal, both to the north-east and to the south-west. The strike of the cleavage (060/70°NW) in the mudstone beds of both limbs can be seen to be up to 10° clockwise to the strike (050°) of the bedding on the limbs, and of the axial surface. But the exposures also allow a profile view of the mudstone in the hinge area; here, the cleavage shows the development of a finite neutral point (Ramsay, 1967 p. 417), with a slightly divergent cleavage fan above it (Figure 2.8B), and a bedding-parallel fabric below. Such features are generally accepted to be the product of strain related to folding, and this example would seem to provide evidence of the contemporaneity of the non-axial plane cleavage and the folding.

Fold B (NX 47984982) is a more open anticline than A, plunging to the south-west. It displays alternating sandstone and mudstone beds with, in profile, classic convergent cleavage fans in the sandstones and near axial-planar cleavage in the mudstones. Because of the plunge and the flat nature of the outcrop, it is possible to walk across successive fold hinges of bedding, along the one axial surface, and it is quite clear that cleavage in both sandstones and mudstones consistently transects the axial surfaces and fold hinges in a clockwise sense. This is one of the most convincing and photogenic (Figure 2.7) exposures that demonstrates this phenomenon.

Fold C (NX 48004992) is also an anticline demonstrably transected by the cleavage, but its principal attraction is the geometry of its hinge area, which is exposed for some twenty metres along its length. The hinge exhibits two gentle plunge culminations, as well as variation in strike of its axial surface. These features are characteristic of many of the small-scale folds of the Southern Uplands.

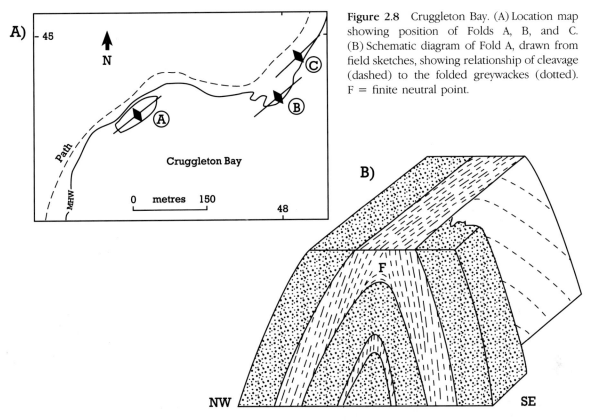

Figure 2.8 Cruggleton Bay. (A) Location map showing position of Folds A, B, and C. (B) Schematic diagram of Fold A, drawn from field sketches, showing relationship of cleavage (dashed) to the folded greywackes (dotted). F = finite neutral point.

Interpretation

Rust (1965) in his description of the deformation history of the Whithorn area proposed that there had been four important phases (F_1–F_4). Stringer and Treagus (1980, 1981) only recognized two phases D_1 (incorporating Rust's F_1 and F_3) and D_2 (Rust's F_2 and F_4). As stated above, these divergent views rest largely on the variation in interpretation of the age of the dominant cleavage in the area, S_1 of Stringer and Treagus, S_3 of Rust. Rust (1965) observed that this cleavage transected folds of his F_1 and F_2 generations, but was axial-planar to locally developed, steeply plunging folds (see Isle of Whithorn) which he therefore regarded as of third generation (his F_3). Stringer and Treagus (1980, 1981), on the other hand, from evidence in this site and other localities, maintain that the dominant cleavage, although transecting, was essentially contemporaneous with the formation of the D_1 folds. These folds locally assume steep plunges, and there they show the axial-planar

relationship of the cleavage in vertical profile view.

The arguments for the contemporary age of cleavage and folding lie in the observation that the cleavage geometry is a product of the strain variations related to the fold geometry. The most commonly observed relationship of this type is the fanning and refraction of cleavage, symmetrical to the axial surfaces of folds in profile view (see Figure 2.6B, Barlocco site). At Cruggleton Bay, the most persuasive relationship is that the clockwise transecting cleavage is clearly seen to have the usual detailed geometry (see Figure 2.8B) around finite neutral points, seen in thickened mudstone hinges between adjacent greywacke beds. Such a relationship is taken to indicate that the cleavage pattern is a direct response to strains that developed in the tightening hinge (Ramsay, 1967, p. 417).

The origin of transecting cleavage, which has now been widely recognized in the Lake District and Wales (Soper *et al.*, 1987), has excited great interest, in the context of the closing of the Iapetus Ocean. Stringer and Treagus (1980, 1981) and Treagus and Treagus (1981), from observations in

Figure 2.7 Cruggleton Bay North. D_1 folds in Silurian siltstones and mudstones, plunging to the south-west, transected by non-axial plane cleavage. (Photo: P. Stringer.)

the Southern Uplands, have suggested that the phenomenon could result from the development of fold axial surfaces in an orientation other than the conventional one perpendicular to the bulk shortening and thus not parallel to the cleavage. The model was developed by assuming that the strain did not depart significantly from plane strain with moderate subvertical stretching, based on the limited field observations of strain parameters. In the Irish equivalent of the Southern Uplands, however, Anderson and Cameron (1979) and Murphy (1985) have recognized transection, again mostly clockwise, but associated with a distinctive, subhorizontal stretching lineation.

Transecting cleavage has been attributed to a deformation model called 'transpression' (see Sanderson and Marchini, 1984), which can be produced by superimposing simple shear on a non-rotational strain. Soper and Hutton (1984) have applied this model to the Caledonides, relating sinistral simple shear, near the Iapetus suture, to the transecting cleavage in Ireland. Similar applications of the model to terranes like the Southern Uplands and Lake District (Soper *et al.*, 1987), where reports suggest that strain is oblate (Stringer and Treagus, 1980) or plane with moderate upward stretching (Bell, 1981), require more detailed examination of strain parameters in rocks such as those at Cruggleton and Barlocco.

Conclusions

The site has been selected for inclusion in the Geological Conservation Review for its illustration of the relationship of cleavage to folding. The three-dimensional forms of a number of large folds can be seen with great clarity at this locality. The cleavage (fine, very closely spaced, parallel fractures) is orientated slightly obliquely to the trend (hinges) of the fold (it is therefore described as being non-axial planar), and yet they are contemporary. The phenomenon is important in the understanding of the movements that ended the Caledonian Orogeny in Britain, around 400 million years before the present. The non-parallelism of fold axial surfaces and cleavage suggests that when the Iapetus Ocean closed at that time, its margins were oblique to the direction of closure.

ISLE OF WHITHORN BAY
(NX 47663650–47603616)
J. E. Treagus

Highlights

A feature of the Caledonian D_1 deformation in the Hawick Rocks of the Southern Uplands is a narrow zone of steeply plunging folds. The Isle of Whithorn Bay is the best locality for the examination of this fold zone.

Introduction

This site occurs within the Hawick Rocks, of probable Llandovery age, in the Central Belt of the Southern Uplands – see 'Introduction', Chapter 1, and Figure 2.9. The regional attitude of these rocks, as described by Craig and Walton (1959), is of steep north-west-dipping and north-west-younging bedded sediments, interrupted by south-east-verging D_1 folds on a variety of scales (Rust, 1965; Stringer and Treagus, 1981). These folds usually display a range in plunge up to 30° to the north-east, or the south-west (see Barlocco and Cruggleton Bay: Figure 2.10).

At Whithorn Bay the plunge variations are on a larger scale and only become apparent from careful examination of successive outcrops. These folds, which are part of a 1500 m wide belt (Figure 2.1) across the Whithorn Peninsula, may be related to one or more strike-parallel thrusts. This site also reveals from examination of way-up criteria, the presence of an isoclinal fold which pre-dates D_1. Such folds have been described from several localities in the Hawick Rocks; they have been interpreted as the products of soft-sediment deformation, developed early in the history of the accretionary prism.

Stringer and Treagus (1980) described this zone, near the south-west margin of the Central Belt (labelled B on Figure 2.1), which exhibits the steeply plunging and locally downward-facing D_1 folds, seen at this site. Rust (1965) interpreted the steep-plunging folds of the Whithorn area as a response to a third episode of shortening, which affected already vertically dipping rocks. Stringer and Treagus (1980) interpreted these folds as an exceptional development of the ubiquitous D_1 folds, with the associated S_1 cleavage.

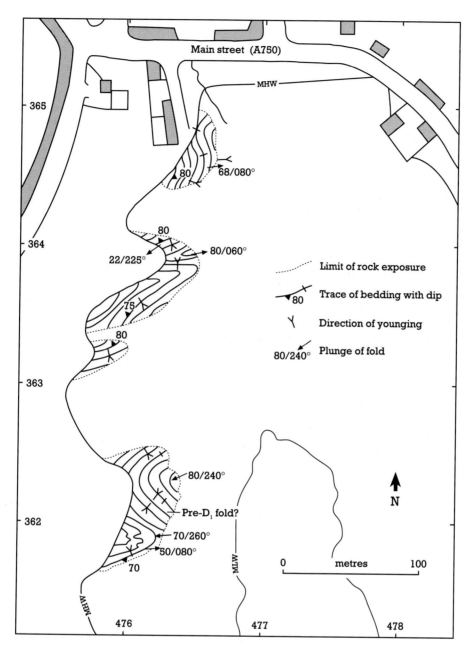

Figure 2.9 Sketch of the Isle of Whithorn Bay site.

Description

The foreshore on the western side of the bay (see Figure 2.9) contains six major D_1 hinges in the typical alternations of greywacke, siltstone, and mudstone. The rocks broadly trend NE–SW, with subvertical dips and younging to the north-west. The short limbs have a similar dip, but more north-westerly strike and young to the south-east. Thus the fold hinges are themselves subvertical and occasionally can be directly observed. In detail, some fold hinges plunge steeply north-east (some as shallow as 40°) and can be traced along their axial surfaces to plunge steeply south-west. Some hinges plunge as shallowly as 20°, giving a total range of plunge of 120° about the vertical. Unlike the John's Road site, it is not possible to demonstrate that individual hinges curve through this range. The variation in plunge is illustrated by Stringer and Treagus (1980, Figure 6c and p. 324).

Figure 2.10 Typical fold plunge variation in Hawick Rock greywackes. This pericline, viewed from the north-west, is at Shaddock Point (NX 478 393), near the Isle of Whithorn site. (Photo: P. Stringer.)

Throughout the site, it is usually possible to demonstrate the direction of younging from sedimentary structures. At the northern end (about NX 47703645), 1 m-thick greywackes display unusually clear examples of graded bedding, with gritty, loaded and flute-casted bottoms, and cross-laminated silts and cleaved mudstones at the tops. Such indications, elsewhere at the site, show that the majority of the folds face sideways, or upwards, to the south-west. The folds that plunge to the north-east can, similarly, be shown to be downward-facing to the south-west. However, a reversal of younging of the rock sequence, at NX 47623622, which cannot be accounted for by the D_1 folds, implies the presence of a major isoclinal fold that pre-dates D_1. Similar pre-D_1 folds at Cairnhead (NX 48673838), considered to be of soft-sediment origin, have been described by Stringer and Treagus (1981, p. 141) and Rust (1965). That the folds are D_1, is clearly demonstrated from the relationship between the folds and the cleavage, the latter is everywhere developed in the finer-grained lithologies. It is approximately axial planar, showing the usual refraction through the various lithologies. The cleavage also exhibits the transecting relationship to the fold hinges, shown throughout the Hawick Rocks (Stringer and Treagus, 1980; Figure 2.11).

Interpretation

The principal interest of this site is that the steeply plunging folds, which it so clearly exhibits, are part of a zone of such structures that runs for some 20 km along the southern margin of the Central Belt (labelled B on Figure 2.1). At Whithorn the zone is 1000 m wide, whereas to the north-east, on the other side of Wigtown Bay, it appears to be represented by two narrower components (Stringer and Treagus, 1980, Figure 1). No descriptions exist of such structures in the poorly exposed inland areas, but the site at Agate Point and John's Road on the north-east coast may be part of the same, or a parallel, zone 200 km along strike.

The origin of the folds has been attributed to a post-D_1 deformation by Rust (1965). This was

Figure 2.11 Typical cleavage refraction from mud/siltstone to sandstone seen in Hawick Rocks in profile view; in plan view cleavage transects the fold hinges. Locality: Port Allen (NX 478 411), near Isle of Whithorn. (Photo: P. Stringer.)

contested by Stringer and Treagus (1980) who pointed out that:

1. these folds were part of a range of D_1 fold plunge in the area;
2. that the folds exhibit, apart from their plunge, all the usual features of D_1 folds in their wavelength, vergence and general style; and that
3. the regionally developed cleavage has the same relationships to these folds as it does to the regional D_1 folds, namely subaxial planar or slightly transecting, and refracting through the various lithologies.

Kemp (1987) described some steeply plunging folds in the Southern Belt in strongly sheared rocks, to the south-east of the zone described here. He showed that the folds have a consistent sinistral vergence and post-date early folds (and presumably cleavage). Clearly, the post-D_1 folds have a different origin, more clearly related to shearing than those discussed here.

The origin of these folds is still unexplained and will undoubtedly be the subject of further research, particularly in view of the unusual deformation characteristics that would be expected as a result of the position of these rocks in a possible accretionary prism. Stringer and Treagus (1980) suggested, firstly, that they might be related to unusual strain gradients which might be associated with the thrusting that is an essential feature of the accretionary process. Folds that curve into the extension direction are usually related to strong extensive strains (Roberts and Sanderson, 1974) often in shear zones (Cobbold and Quinquis, 1980). However, no exceptional strain parameters have been reported from these rocks, indeed the fabric suggests oblate strain. Stringer and Treagus (1980), secondly, suggested that D_1 folds may have been rotated in packets between shear planes as part of the thrust related (D_1 and later?) deformation. The boundary between the Hawick Rocks and Wenlock strata, immediately to the south-east of the zone (see West Burrow Head), may be the location of one such thrust. The site certainly contains planes or zones along which there is local disruption, intensification of cleavage and veining. These zones could mark the boundaries of anastomosing minor thrust packets, although there is apparently no great discontinuity of fold structure, or lithological type across them. The age, the sense of shear, and the relation to external stresses of the zone of steeply plunging

folds is an obvious target for future work.

A second feature of interest here is the apparent presence of a pre-D_1 isoclinal fold. Such folds, unrelated to cleavage, or minor structures, have been attributed (Stringer and Treagus, 1980, 1981) to soft-sediment deformation. Again, this feature, as well as those of similar origin described by Knipe and Needham (1986) and Kemp (1987), need to be further studied. Soft-sediment structures need to be more closely related to those in accretionary prisms. Their geometry and origin in the Southern Uplands or in modern prisms cannot yet be related with confidence.

Conclusions

The site has been included in the Geological Conservation Review as the most convincing and accessible location for the study of steeply-plunging folds in the Southern Uplands. These folds, which vary considerably in their orientation (plunge), through a total range of 120° about the vertical, are highly unusual features of slate belts and must be an important, but as yet poorly understood, feature of the accretionary development of the north-western margin of the Iapetus Ocean. In the Hawick Rocks of this area they are characteristic. These folds were the product of extreme compression during the Caledonian mountain building episode at the end of the Silurian Period or early in the Devonian. They affect older folds here, thought to have formed by the slumping of unconsolidated sediments on a sloping early Silurian sea-bed, perhaps initiated by disturbances in the early stages of accretion. Thus the locality displays graphically sedimentary and tectonic deformation over a period of around 40 million years.

WEST BURROW HEAD (NX 45183411–45343411)
J. E. Treagus

Highlights

This site has total exposure across the stratigraphically and structurally important boundary between the Hawick Rocks and fossiliferous strata of the Lower Wenlock Series. The boundary here is clearly faulted, but the throw of the fault has been the subject of controversy.

West Burrow Head

Introduction

This coastal locality is important in the controversy concerning the nature of the junction between the belt of Wenlock rocks, in the south-east, and the Hawick Rocks, to their north (see Walton *in* Craig, 1983 p. 129). At Fouldbog Bay, the along-strike equivalent of this steeply dipping junction is claimed, by Craig and Walton (1959), to be marked by transitional beds from the younger Hawick Rocks south-eastwards to the Wenlock. They concluded (see also Clarkson *et al.*, 1975) that the Hawick Rocks must therefore have either a latest Wenlock or a Ludlow age. At Burrow Head, Rust (1965) claimed that transitional rocks were absent and that the junction exposed between the Hawick Rocks (assumed by him to be Upper Llandovery) and the Wenlock, was a fault which required a southward downthrow of some 3 000 m. Rust considered that the apparent transitional rocks (red mudstones associated with graptolitic Wenlock) might have been produced by fault slicing at Burrow Head, Fouldbog Bay and elsewhere.

Most recently, Kemp and White (1985) and Kemp (1986), working in the Fouldbog area and in areas further to the north-east, has claimed that the Wenlock strata south-east of the Hawick Rocks are an intensely imbricated sequence of packets successively younger in that direction; like Rust (1965), he has attributed a Late Llandovery age to the Hawick Rocks. Barnes *et al.* (1987, Figure 4) confirm Rust's (1965, Figure 4B) observation, and that of the present work, that the site lies on the short limb of a major SE-verging fold pair, with a wavelength of about 1 km. Barnes (1989) cites the graptolite evidence which firmly assigns the rocks here to the Llandovery.

Description

At the north-west end of the site (Figure 2.12A and B), regularly bedded, south-east dipping, Hawick greywackes and siltstones young consistently upwards, interrupted rarely by fold pairs verging to the north-west. Two distinctive red mudstone beds occur near the top of the exposed succession (Figure 2.12A). Some 20 m to the east of the red mudstones, a zone, about 10 m wide, is traversed by five steep fracture planes trending north-east (Figure 2.12B). The fractures are not associated with any of the usual fault-plane features suggesting major movement, except possibly thin mylonitic banding. The slices between the fractures usually retain some coherence of bedding, although a 1 m-wide central slice contains strikingly lensoid beds of greywacke. One syncline and one anticline core are evident elsewhere in the slices. To the south of the fractures, the first coherent greywackes, again clearly young to the south-east, but are immediately involved in a tight syncline. The north-west-younging continues through beds containing graptolitic mudstones of Wenlock age (see Rust, 1965, Figure 2), but is reversed again to the south-east after 20 m by the complementary anticline. The Wenlock rocks continue to dip and young to the south-east for at least another 200 m towards Burrow Head, with rare interruption by north-west-verging fold pairs. 2.12A & B

Interpretation

This site demonstrates that the junction between the Wenlock and Hawick Rocks, at least locally, is a fault zone. Apart from the physical evidence of fracture planes separating the two units, there is no match of lithology across the zone. As the sense of shear and the precise stratigraphical separation across the zone are as yet unknown, its significance can only be speculated on from regional considerations. Detailed structural investigation of the site itself and of the rocks immediately north and south could provide an answer to the debate as to whether the Hawick Rocks are younger (Craig and Walton 1959) or older (Rust, 1965) than the Wenlock.

Although recent opinion favours the view that the Hawick Rocks are attributable to the Llandovery Series (Kemp, 1986; Kemp and White, 1985; Barnes *et al.*, 1987), the opposing view still has the merit that it requires the least displacement on the fault zone at Burrow Head. Thus, if the red mudstone lithologies in the greywacke of the local Hawick Rocks were transitional downward to the Wenlock, as Craig and Walton (1959) propose, then the Wenlock would lie hidden in the core of the anticline to the north of the fault zone illustrated in Figures 2.12A and 2.12B. The red mudstones, to the south of the fault, would lie above the Wenlock in the core of the synclinal complex which lies south of the site. Displacement on the fault zone might then be a few hundred metres, but downwards on the north side.

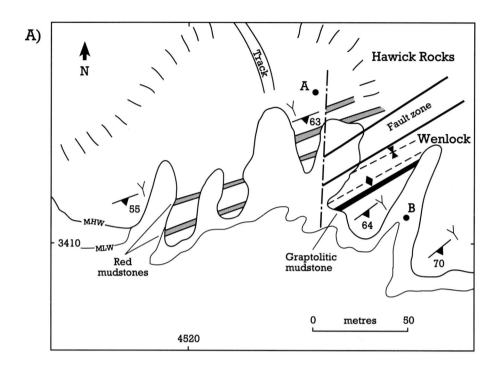

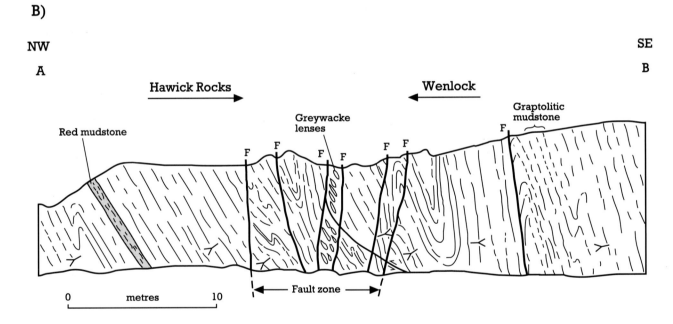

Figure 2.12 West Burrow Head. (A) Geological map of site. (B) Sketch cross-section along line A–B of Figure 2.12A).

On the other hand, if Rust (1965) is correct in his view that north of the anticline (north-west of the site) there is a sequence of Hawick Rocks (Rust's Carghidown and Kirkmaiden Formations) uniformly dipping and younging to the north-west as far as Kirkmaiden, then a 2–3 km displacement would be necessary, downwards on the south side of the fault zone shown in Figures 2.12A and 2.12B. The magnitude of the displacement, in the latter hypothesis, depends very much on the stratigraphical and structural arrangement of the Carghidown and Kirkmaiden Formations. Information relating to this is given in the Geological Survey Memoir (Barnes, 1989). Barnes *et al.* (1987) suggest that the Carghidown Formation (with red-beds) is younger than the Kirkmaiden Formation (with a gradational boundary between them) and may be stratigraphically overlain by the Wenlock rocks at Burrow Head. The structural arrangement, whereby this stratigraphy is obtained, is not yet published, but must involve a new interpretation of folding and/or faulting in the Hawick Rocks.

Thus the Burrow Head site is potentially of great importance, in the resolution of two issues in Southern Upland geology. Firstly, there is the question of the age of the Hawick Rocks and all that it implies for the palaeogeography. If the interpretation of a transitional junction is correct then only a small displacement, down on the north side of the fault zone, would be required. Such a fault would not be a major down-to-the-south thrust associated with the accretionary process. On the other hand, a smaller displacement, possibly of D_1 age would explain the lack of high-strain features associated with the fault zone. Secondly, there is the question of the position, age and displacement of the faults which are critical to the interpretation of the structural evolution.

Conclusions

This site is an example of one of the major, steep, strike-parallel fault lines that form an important part of the structural framework of the Southern Uplands. It has been claimed that the fault has a very large displacement of 3000 m, bringing mid-Silurian aged rocks (Wenlock) against early Silurian Hawick Rocks; this dating is based on sparse fossils found in these marine sediments. The displacement of the fault depends on the precise dating of the rocks on its two sides, and an alternative 'structural' interpretation has been proposed which would make the fault a far less important structure. The settlement of the controversy over this major Southern Uplands discontinuity, will have considerable repercussions on the study of the geological framework of southern Scotland.

BACK BAY, MONREITH (NX 36783947–36943930) J. E. Treagus

Highlights

Back Bay, Monreith has the most dramatically exposed cliff section of large-scale folding in the Southern Uplands (Figure 2.13). Two large-scale, D_1 fold pairs are completely exposed in the 30 m-high cliffs, with an excellent example of D_2 structures superimposed. The geometry of these features is characteristic of the deformation style in the Central Belt of the south-west Southern Uplands.

Introduction

The rocks of this site display a typical range of lithologies in the Hawick Rocks, including some massive greywackes, as well as thinner-bedded greywackes, siltstones, and mudstones. The principal feature of the site is the refolding of the north-westerly fold pairs by a D_2 fold pair with a flat-lying south-easterly dipping axial surface and crenulation cleavage. The style of this D_2 deformation is typical of its local development throughout the Central Belt, but here is part of a 2.3 km-wide zone that crosses the Whithorn Peninsula and can be traced to the north-east coast of Wigtown Bay (Figure 2.1). The D_2 deformation is locally associated here with minor north-west directed thrusting, but regionally appears to result from subvertical shortening of the D_1 fold stack.

The site is part of the Whithorn area described by Rust (1965) in which he recognized four phases of deformation related to folds and cleavage. The F_1 folds of Rust are the D_1 folds of the West Burrow Head site 10 km to the south, and of the present description. Superimposed on this phase Rust recognized two sets of folds, his F_2 and F_4

Figure 2.13 Type-D_1 syncline in massive Silurian greywackes, Back Bay, Monreith. View to the north-east, with figure for scale. (Photo: J. Treagus.)

phases, and the Back Bay locality is quoted (Rust, 1965, p. 14 and Plate 2) as an example of the F_4 folding with accompanying cleavage. The distinction of F_2 and F_4 appears to be partly on the flatter dips of the axial surfaces of the latter and also on the superimposition of F_4 on the locally developed, steeply plunging F_3 folds.

Stringer and Treagus (1980, 1981), recognized one dominant phase of deformation, D_1, with its associated (but non-axial planar) cleavage. This is locally affected by a D_2 deformation (incorporating Rust's F_2 and F_4) particularly in one 2 km-wide, NE-trending belt (labelled A on Figure 2.1), on which the Back Bay site lies in a median position.

Description

The north-western half of the section (Figure 2.14) is dominated by a spectacular profile of a large open D_2 fold (minimum amplitude 5 m) super-imposed upon a tight upright D_1 anticline (at NX 36823942) and a syncline (at NX 36853940) of *c.* 30 m amplitude (Rust, 1965, Figure 1 and Plate 2). The folds plunge gently to the north-east. At the north-west end of the section, steep NW-dipping strata and S_1 cleavage are deformed by smaller-scale, D_2 'step' folds with axial surfaces and sporadic, D_2 crenulation cleavage inclined gently to the south-east (Stringer and Treagus, 1981, Figure 3A). Minor thrust faults, with small north-westerly displacements and a zone of quartz veins, are associated with the D_2 folds.

The south-eastern half of the section (Figure 2.14) comprises a D_1 anticline (at NX 36913937) and syncline (at NX 36933936, see Figure 2.13) forming an asymmetrical D_1 fold pair (Folds 3 and 4 in Stringer and Treagus, 1981, Figure 3A) which verges to the south-east. The folds are tight (interlimb angle 20–35°); overturned strata in the common short limb (15 m wide, measured horizontally) dip steeply to the north-west, and strata in the long limbs (50 m wide) dip moderately to steeply to the north-west. The D_1 fold pair

NW   SE

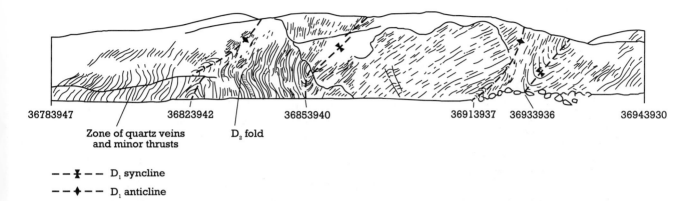

| 36783947 | 36823942 | 36853940 | 36913937 36933936 | 36943930 |

Zone of quartz veins D_2 fold
and minor thrusts

— — ⚹ — — D_1 syncline

— — ✦ — — D_1 anticline

Figure 2.14 Sketch of the Back Bay site, drawn from photographs. Horizontal scale is not linear; total length is approximately 200 m.

(wavelength *c.* 75 m, amplitude *c.* 30 m) is probably intermediate in scale to larger scale, 0.25–3.0 km, D_1 folds. Gentle curvature of the north-westerly-dipping axial surfaces indicates large-scale D_2 folding. The D_1 cleavage in mudstone and sandstone beds strikes close to 070° and 090° respectively, markedly oblique to the north-easterly strike of the D_1 fold axial surfaces.

The site generally offers the opportunity to examine all aspects of the D_1 and D_2 deformations, including the relative thickening of fold hinges compared with limbs, boudinage of limbs, the variable development of the S_2 cleavage, and the associated quartz veining and thrusting.

Interpretation

The interest of the site lies in the scale and geometry of both D_1 and D_2 folds, and the fact that it affords the rare opportunity to observe one phase superimposed on the other.

The D_1 folds give a glimpse of the amplitude and wavelength of this phase of folding, although the first-order D_1 folds in the Central Belt are thought by Stringer and Treagus (1983) to be of the order of 0.25–3.0 km. The section (see Figure 2.14) also reveals the south-east vergence of these second-order folds, as seen by the relative thickness of the north-west- and south-east-younging limbs respectively. In fact, the north-west-younging limb, between the north-west and south-east fold pairs, is uncharacteristically gently dipping, pre-

sumably as a consequence of D_2 deformation. A 30° sheet-dip to the north-west would probably represent the pre-D_2 value, seen here interrupted by major folding. In the accretionary model, these folds would be interpreted as having a primary attitude with gentle landward-dipping (NW) axial planes and south-east vergence. Their rotation to their approximate present attitude would have taken place during thrusting, possibly exemplified at the West Burrow Head site. Finally, the transecting cleavage would have been impressed on the folds during the late Silurian closure of the Iapetus and before dykes were emplaced, the latter dated by Rock *et al.* (1986) at 418–395 Ma.

The D_2 folding shows the typical (but often steeper) dip to the south-east of the axial surfaces of the folds and their consequent south-easterly vergence. Like Rust (1965), but for different reasons, Knipe and Needham (1986), elsewhere, distinguished two generations of folds with this geometry. The earlier, with steeper axial surfaces, they ascribe to pre-accretion rotation and associate with continuing oceanward rotation from D_1 (Knipe and Needham, 1986, Figure 11). The second set, with flatter, south-easterly-dipping axial surfaces like those at Back Bay, they attribute to post-accretion rotation and note the association with minor thrusting directed to the north-west. They associate this structure with shortening in the collision related to the closure of the Iapetus. Stringer and Treagus (1980, 1981) were unable to distinguish these two sets of post-D_1 structures, either on the basis of interference of one set with another or from differences in morphology or mineral growth related to the respective crenulation cleavages. The present description shows that all structures here called D_2 must post-date the S_1 cleavage and must reflect very late (post-395 Ma) post-collision adjustments, possibly related to the adjacent granite bodies of this age. The Back Bay site would be an obvious locality for further work to unravel this contentious post-D_1 deformation history.

Conclusions

This site is important as the best-known locality where the Caledonian D_2 folds of the Southern Uplands can be clearly seen to be superimposed D_1 folds, formed earlier in that mountain-building phase. Large, upright asymmetrical folds with amplitudes up to 30 m, associated with steeply inclined cleavage (fine, closely spaced, parallel fractures), are refolded by a second generation of smaller folds orientated almost at right-angles to the first. The first generation (D_1) of folds is thought to result from the subduction of oceanic crust during the closure of the Iapetus Ocean, whereas the cleavage probably relates to eventual continental collision in the end-Caledonian mountain-building episode. The second generation of folds (D_2) is possibly related to uplift. As well as providing an unequalled view of the style and scale of D_1 folds, this locality must rank as one of the most dramatic large-scale exposures of major refolded folds anywhere in the British Isles.

GRENNAN BAY (NX 07754373)
J. E. Treagus

Highlights

This site exposes one of the 'inliers' of Moffat Shales within the Llandovery greywackes of the Central Belt. Here the contacts between the two units are unusually well displayed. The Moffat Shale inliers have played a central role in the debate over the structure and geological history of the Southern Uplands, for more than a century.

Introduction

The Central Belt of the Southern Uplands, like

the Northern and Southern Belts, comprises essentially steeply dipping and dominantly north-west-younging greywackes. In addition to the south-east-younging sequence of the three belts (Ordovician, Llandovery, and Wenlock in the main), the greywackes within the Central Belt also exhibit a sequence of blocks decreasing in age to the south-east – see 'Introduction' to this chapter, Figure 2.2C and Leggett *et al.* (1979). Craig and Walton (1959) and Walton (1961) proposed that these blocks must be separated by steep strike faults with downthrows on their south-eastern sides (Figure 2.2B). The evidence for these faults comes largely from the presence, within the poorly fossiliferous greywackes (of generally Late Llandovery age), of the graptolitic Moffat Shales, (Hartfell, Birkhill, and Glenkiln Shales, of Llandeilo to Llandovery age). The juxtaposition of Moffat Shale slices above, and to the north-west of, the greywackes in the Grennan Bay site requires faults with substantial throws. Lapworth (1889) and Peach and Horne (1899) in the earliest interpretation of the Moffat Shale strips within the greywacke, saw them as anticlinal inliers of the Ordovician, although they fully appreciated their faulted nature (Figure 2.2A). This site is important in its bearing on the structural relationships between the Moffat Shale 'inliers' and the younger Llandovery greywackes of the Central Belt. At its southern margin, this site exposes a gradational contact between the Moffat Shales and these greywackes to their north. The northern half of the site consists of a 100 m-thick slice of Moffat Shales, with fault contacts with the greywackes to their north and south (Figure 2.15). The latter fault is an example of one of the steep 'thrusts' which, in the accretionary interpretation of the Southern Uplands, would be consequent to the north-westerly subduction of the Iapetus Ocean crust.

The area of the Rhinns of Galloway in which the Grennan Bay site lies, had not been studied since the original Geological Survey mapping (Peach and Horne, 1899). However, it has recently been the subject of resurvey by J. A. McCurry (Barnes *et al.*, 1987), who provided much of the background information for the present description.

Description

The stratigraphy at Grennan Bay, together with post-cleavage lamprophyre dykes and faults, is demonstrated at eight localities within the site (a–h; Figure 2.15); these are discussed in detail

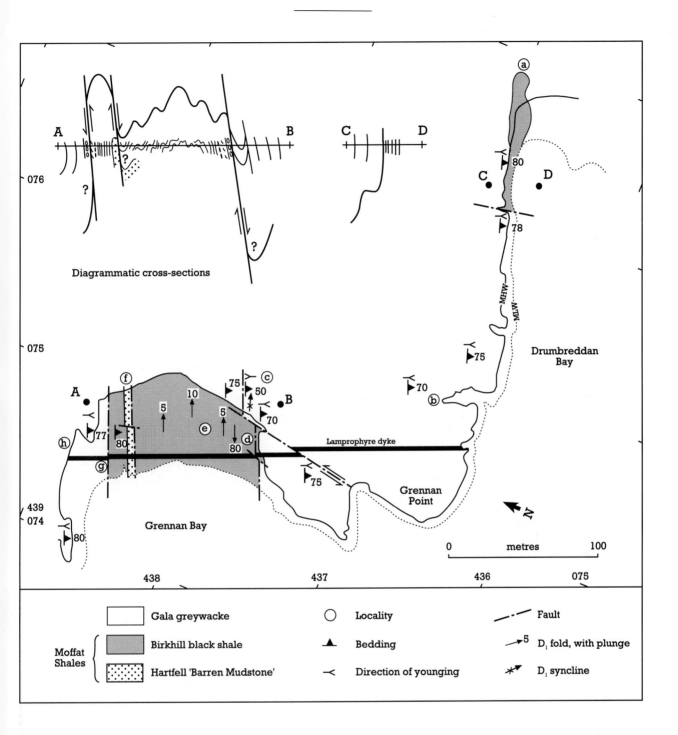

Figure 2.15 Geological map and cross-sections (inset) of the Grennan Bay site (after J. A. McCurry, unpublished).

below. The structure is illustrated by the cross-section in Figure 2.15.

a) The southern boundary between the Moffat Shales and the greywackes displays transitional lithologies between the grey/black Birkhill Shales (Llandovery) and the grey shale and siltstone lithologies in the greywackes (late Llandovery) to the north. This provides an ideal contrast with the two boundaries (d and g) further north, which are clearly faulted and where these transitional lithologies are absent. Dips are subvertical at the contact, and there is north-west younging in both the transitional lithologies and in the greywackes.

b) The southern unit of greywackes consists of 1 to 3 m-thick greywackes with subordinate grey siltstone and mudstone at the north-west margins of graded units. The beds dip steeply to the south-east, apparently unaffected by D_1 folds; they are inverted and thus young consistently to the north-west.

c) The northern twenty metres of the greywacke is affected by a faulted syncline, some 10 m across, clearly seen on the eastern wall of a later NNE-trending fault which also cuts the boundary with the Moffat Shales. The southern limb of this syncline is interrupted by a steep, narrow fault zone. The movement on this fault may well be contemporary with that on the Moffat Shale margins, mentioned below. No complementary anticline is seen before the Moffat Shales junction (d) is reached, although unequivocal south-easterly-younging was not found in the greywacke at the junction.

d) The greywacke at the southern junction with the Moffat Shales block is, therefore, assumed to be right way-up, dipping some 60°SE. The first metre of the Moffat Shales, north of the immediate junction, is strongly disturbed by a fault zone, the effects of which extend for a further three to five metres into the Moffat Shales to the north. Within this first metre, lenses of greywacke, up to a metre in length, occur in sheared black shale and the Moffat Shales beyond are broken by anastomosing fracture planes. The actual junction with the Moffat Shales is generally sharp and planar (dipping steeply to the south-east), but in detail it is often undulating. Some striations are subvertical within the fault zone, but there is no other obvious indication of direction, or amount of movement. Some movement disturbs, and thus post-dates, the S_1 cleavage. The simple

structure of the greywackes (c) and the position of the Moffat Shales (a) and (e), suggest that movement on this fault (assuming it to be subvertical) was down on its south-east side.

e) The Hartfell Shales have yielded Caradoc graptolites (*C. wilsoni–D. clingani*) and the Birkhill Shales have yielded a Lower Llandovery fauna (*M. convolutus* biozone). The Moffat Shales are well bedded, away from the faulted margins, but do not readily reveal their 'way-up'. They are interrupted by several zones, a few metres wide, of D_1 folding. These folds, with a wavelength of 0.5–1 m, generally plunge at gentle angles to the north-east, but also exhibit plunges up to 50°NE and 80°SW. The vergence of the folds, in the central outcrops, appears to be neutral, or to the south-east, and this supports the fragmentary evidence of north-westerly younging (assuming that the folds face up). It is possible that the south-east part of the slice of Moffat Shales could be on the southern, south-east-younging, limb of an anticline (as in the cross-sections of Peach and Horne and in Figure 2.15 herein) interrupted by the southern fault (at point d) and separated from the succeeding greywackes; this is supported by the observation (McCurry, pers. comm.) that the southernmost Moffat Shales contain common bentonites, suggesting they represent the youngest part of the Birkhill Shales. Strike-parallel faults (of unknown displacement) are certainly present within the Moffat Shales slice, making the reconstruction of fold geometry more difficult. The S_1 cleavage is only locally developed in the Moffat Shales in the axial zones of the D_1 folds; it appears to be subaxial planar.

f) Towards the northern end of the main outcrop of Moffat Shales is a zone of grey mudstone/shale, some 10 m wide, which is presumed to be part of the upper Hartfell 'Barren Mudstone'; the northern junction with the Moffat Shales appears to be a sedimentary and conformable junction younging northwards, but the southern junction is probably a fault. Since the southerly junction is followed by a syncline in the Moffat Shales, the outcrop of the Barren Mudstone may well be anticlinal with a thinned and faulted southern limb, as shown in Figure 2.15.

g) The junction of the Moffat Shales with the greywackes to the north is, again, clearly a fault zone. The actual junction was only seen in one place (but this is dependent on the movements of material on the foreshore), over a distance of 1 m, where it is a sharp plane, steeply dipping to the south-east, between black shale and platy grey siltstone. The Moffat Shales are broken by fractures, and bedding is lost within a zone some 3 m-wide to the south of the junction. To the north, thin beds of greywacke display excellent lensoid structure in a matrix of platy siltstone and grey shale for several metres; D_1 folds have been disrupted. Again, subvertical striations provide a suggestion of the movement direction. If the movement direction is subvertical then the stratigraphical separation would require displacement down on its north-west side.

h) The greywacke beds to the north of the fault zone are 0.5–2 m thick, dipping steeply to the south-east. Within ten metres of the fault zone they are inverted and show uniform north-west younging.

The Moffat Shales 'inliers' are of great importance in studies of the Caledonian structural history of the Southern Uplands. They are not simple anticlinal inliers as envisaged initially by Lapworth (1889), but have come to be seen, from the work of Walton (1961), Toghill (1970), and, most recently, Webb (1983), as the expression of D_1 fold pairs modified by faulting, particularly on their SE-facing short limbs. Their stratigraphical and structural relationships suggest that each is marked by one, or more, powerful faults downthrowing to the south-east ('thrusts' if dipping to the north-west), such that Moffat Shales in the Central Belt are repeatedly brought down to the surface. More importantly, such faults, cumulatively, are believed to more than counteract the effect of the steep north-westerly sheet-dip of the Lower Palaeozoic rocks of the Southern Uplands, so that successively younger rocks are encountered in a south-easterly direction. These faults have further come to be seen as the effect of an oceanic plate being subducted, in a north-westerly direction, beneath an oceanic trench, causing the sediments to be sliced and accreted on to the forearc prism to the north-west (Figure 2.2C and Leggett *et al.*, 1979, 1983).

The Grennan Bay site offers clean, almost continuous exposure across three junctions in two inliers. It is superior, in this respect, to any of the exposures of the inliers that have been described already (for instance, Toghill, 1970; Fyfe and Weir, 1976). Although the displacement on the bounding faults are yet to be proved and the internal structure and stratigraphy of the Moffat Shales has not been mapped in detail, the potential for future study at this site is very great (it is currently being studied by J. McCurry, Aberdeen). What the site clearly demonstrates is that here, at least, the outcrop of the Moffat Shales is a consequence of a combination of north-west younging, tight folding and complex faulting. It strongly suggests that a D_1 fold pair is present and that the faults are closely related to the D_1 folds and subparallel to the axial surfaces and S_1 cleavage associated with those folds. These folds are interpreted as examples of the south-easterly verging fold pairs that characterize the Southern Uplands, but here the incompetent Moffat Shales have acted as a detachment horizon for at least one substantial (and now vertical) fault, in part contemporaneous with the folding and/or cleavage. In the accretionary model this relationship would be interpreted to demonstrate the continuity of the north-west underthrusting of the oceanic crust, producing initially south-easterly-verging flat-lying folds which were subsequently detached at Moffat Shales horizons; thrusts and folds were then rotated to their present attitude during accretion. The S_1 cleavage should be the last imprint of the Iapetus closure, and may well have been associated with later strike-dip movement on the thrust planes. In detail, the S_1 cleavage appears to be disturbed by the faulting, but it is thought that this reflects a continuing history of movement that has taken place in the fault zones. Although one fault, on the southern margin of this inlier, certainly has a powerful downward displacement to the southeast (a steep normal fault), another fault at the northern margin must have a substantial downward displacement to the north-west. More work needs to be done, in particular to determine the sense of shear on the fault planes and fault zones.

Conclusions

This locality affords outstanding opportunities to study sections in one of the Moffat Shales 'inliers' of the Southern Uplands, with faulted and unfaulted contacts between the Upper Ordovician to lowermost Silurian Moffat Shales and the surrounding Lower Silurian greywackes. These inliers (that is, areas of older rocks surrounded by younger ones) are now regarded as first-generation (D_1) folds, formed during the early stages of Caledonian mountain building and modified by later faulting. Both the folding and faulting are thought to result from the descent of ocean crust beneath the continent (subduction) on the north-west margin of the Iapetus Ocean: a situation similar to that seen today just east of Japan where the Pacific oceanic plate plunges beneath the Asian continental plate. When this happened in the Southern Uplands during late Silurian to early Devonian times the rocks and sediments were added (accreted) on to the continental margin in a sliced (faulted) wedge or prism.

Chapter 3

Lake District

Introduction

INTRODUCTION

K. Fraser, F. Moseley and J. E. Treagus

A structural perspective

The Lake District of north-west England consists of a major Lower Palaeozoic inlier ($\sim$2600 km^2) surrounded by Upper Palaeozoic rocks. Three major stratigraphical divisions have been identified, which generally young from north-west to south-east (Figure 3.1):

1. The oldest rocks (Ordovician; Upper Tremadoc–Upper Llanvirn) form the Skiddaw Group, a sequence, *c.* 4000 m thick, of mudstones, siltstones and turbiditic sandstone (Wadge, 1978a). These are overlain in the north and east by the Eycott Group (Lower Llanvirn) (Downie and Soper, 1972) consisting of tholeiitic volcanic rocks and interbedded volcanic and sedimentary sequences, *c.* 2500 m thick.
2. The Borrowdale Volcanic Group (Ordovician; mainly Caradoc Series) unconformably overlies the Skiddaw Group and shales correlated with the Eycott Group. This group comprises *c.* 6000 m of calc-alkaline lavas, sills, and pyroclastic deposits, believed to have been erupted under subaerial conditions.
3. The Borrowdale Volcanic Group is, in turn, unconformably overlain by the Upper Ordovician–Silurian (Ashgill–Pridoli) Windermere Group consisting of *c.* 4500 m of shallow-water clastic and carbonate sequences, graptolitic mudrocks, siliciclastic turbidites, and siltstones. These are post-dated by the molasse-type Mell Fell Conglomerate of probable Devonian age (Wadge, 1978b).

In detail, a regional lithostratigraphical framework has been slow to emerge, owing to a paucity of palaeontological control and the complex structure of the rocks.

The rocks of the Lake District were highly deformed (folded, cleaved, faulted) during the major, early-Palaeozoic, late-Caledonian (Acadian) Orogenic event. Several deformation events have been identified, the main one (D$_1$) reflecting the final stages of continent–continent collision. The characteristic Caledonoid NE–SW-trending grain of Lake District structures was developed during this main event. Each of the major rock groups, outlined above, exhibits a distinctive tectonic style. The differences can be related to two main factors: strain history and lithology.

The Windermere and Skiddaw Groups comprise very similar lithologies, but the latter reveals a far more complex structure. The Ordovician Skiddaw and Borrowdale Volcanic Groups are affected by several events which have been related to the final closing stages of the Iapetus Ocean, as well as the eventual continental collision. On the other hand, the dominant structures within the Windermere Group, can be attributed to deformation associated with continent–continent collision.

Lithological differences appear to have exerted a very strong control on how the rocks behaved during deformation. Of the Skiddaw and Windermere Group sedimentary rocks, interbedded mudrocks and sandstones (for example, turbidite sequences) display the best examples of fold structures in the Lake District, while the shales and mudstones display smaller-scale folds of greater complexity. The massive arenaceous sequences and the Borrowdale Volcanics are characterized by larger-scale open folds.

Early structural studies of the Lake District rocks were carried out by Marr (1916), Aveline (1872), and Green (1915, 1920). The substantial and often controversial results of the early studies have been summarized in reviews by Hollingworth (1955) and Mitchell (1956a). In the last two decades significant progress has been made in understanding the Lake District geology and a vast amount of work has been reported in the literature. The recent interest in this area can be traced back to the mid- to late-1960s. Simpson (1967) was the first to apply a modern structural analysis to the rocks of the Skiddaw Group and to observe that they had undergone polyphase deformation (Table 3.1). It was also realized that the Lake District rocks recorded valuable information regarding the final stages of closure of the Iapetus Ocean, in early Palaeozoic times (Wilson, 1966; Dewey, 1969).

The Lake District lies to the south of the postulated suture line between Laurentia to the north-west, and Avalonia to the south-east (Soper *et al.*, 1987); this suture is believed to lie along the Solway line (Moseley, 1977). The consensus is that the Lake District was part of the continental margin, on the northern flank of Avalonia, during the Ordovician and the Silurian, when the initial closing of the Iapetus and the gentle encroachment of the two continents was taking place (McKerrow and Soper, 1989; Fortey, 1989). For example, the observed progressive north to south variation in the Ordovician volcanic rocks, from the tholeiitic tendencies of the Eycott Group in the north of the Lake District, to the calc-alkaline Borrowdale

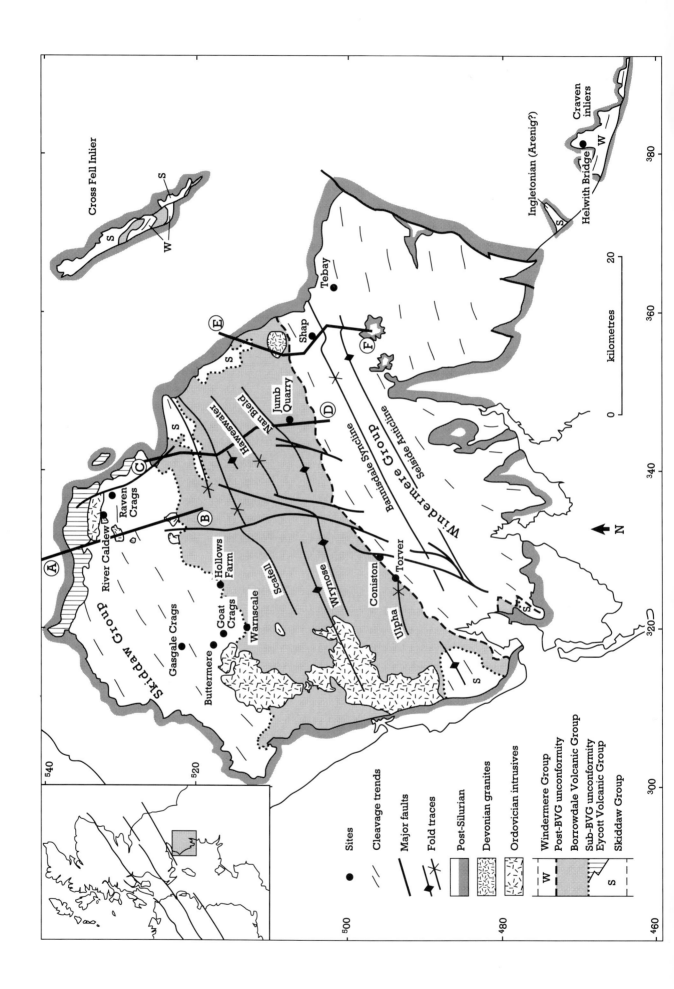

Group (and the predominantly alkali volcanics of Wales) to the south, has been related to a southerly or south-easterly inclined subduction zone at this time (Fitton and Hughes, 1970). The major deformation of the Lake District, however, took place as a result of continental collision. This event, associated with much crustal shortening, regional metamorphism and the intrusion of granitic batholiths in the Lake District rocks, occurred in Early Devonian times (Soper *et al.*, 1987).

It has been suggested (Moseley, 1972) that the late-Caledonian Orogeny in the Lake District can be roughly divided into three phases of: pre-Borrowdale Volcanic Group, pre-Windermere Group, and Early Devonian. The first two phases may be related to subduction and the closure of the Iapetus and the third, and main phase, to the resulting continental collision (Table 3.1).

Deformation phases

The following sections outline the characteristic structures believed to relate to each phase, and the problems with and controversies over their interpretation. Table 3.1 outlines various proposed deformation schemes for the rocks of the Lake District. These are discussed in the text below; the one adopted here is that developed by Soper (1970) and Moseley (1972), with a modification as a result of recent work by Webb and Cooper (1988) and Branney and Soper (1988).

Pre-Borrowdale Volcanic Group deformation

The earliest events occurred prior to deposition of the Borrowdale Volcanic Group; the effects of these may be observed in structures affecting the Skiddaw Group. Simpson (1967) identified three

periods of deformation: F_1, F_2, and F_3 as outlined in Table 3.1. He proposed that the Skiddaw Group–Borrowdale Volcanic Group junction was an unconformity of orogenic proportions, with intense folding (F_1 and F_2) and cleavage development before the start of volcanicity, and only the F_3 deformation affecting later groups. The geometry of Simpson's (1967) F_1–F_3 folds, is essentially that recognized now, for the three deformations, D_1–D_3, affecting all three groups. However, Soper (1970) observed that at all localities where the junction is exposed, there is one prominent ENE cleavage in the Skiddaw slates (S_1 of Simpson, 1967) which passes into the volcanic tuffs above, and he therefore believed the junction to be essentially conformable, with major orogenesis coming after both the Borrowdale Volcanic Group and the Windermere Group. These distinctive views sparked off a heated controversy in the early 1970s (detailed in Moseley, 1972 and discussion; Soper and Moseley, 1978).

Although a number of authors supported the hypothesis that the junction represented a major unconformity (for instance, Helm, 1970; Helm and Roberts, 1971; Helm and Siddans, 1971); Soper and Roberts (1971) substantiated Soper's earlier conviction (1970) by demonstrating that andalusite crystals in the aureole to the Skiddaw Granite (Early Devonian in age) were deformed by the F_2 and S_2 of Simpson (1967): thus demonstrating that F_2 and the closely associated F_1 could not conceivably be pre-Borrowdale Volcanic Group in age, rather they are late-Caledonian structures. It has been shown, however, from the variable age of formations within the Skiddaw Group immediately below the volcanics (Arenig in the west, to Late Llanvirn in the east) that an unconformity does exist, although it is no longer believed to be consistent with orogenesis (Jeans, 1972; Roberts, 1971; Wadge, 1972; Webb, 1972; Moseley, 1972; Soper and Moseley, 1978).

There was general agreement that all the deformation events mentioned (F_1–F_3 of Simpson, 1967) were related to late-Caledonian movements, but doubt remained as to whether there were any structures in the Skiddaw Group which could be attributed to a pre-Borrowdale phase.

Several workers have noted that the NE- and ENE-trending folds, associated with the main cleavage (F_1 and S_1 of Simpson, 1967) in the Skiddaw Group, are superimposed on, and modify earlier northerly-trending folds with no associated cleavage (Roberts, 1971, 1977a; Jeans, 1972; Webb,

---

Figure 3.1 Geological map of the Lake District, and Cross Fell and Craven Inliers, showing lithostratigraphical groups, and major folds and faults of Caledonian age (adapted from Moseley, 1972; Branney and Soper, 1988).

Table 3.1 Deformation sequences in the Lake District as interpreted by various authors; the last column shows the system adopted in the present volume

Stratigraphy and timing of events	Description of deformation phase	Phase numbering and contributions by various workers					
		Simpson (1967)	Soper (1970) and others (see text)	Moseley (1972)	Roberts (1977)	Webb and Cooper (1988)	This volume
	FAULTING dominantly N and NW trends						
	N-S FLEXURES with weak fracture cleavage				D_4		D_3
	RECLINED FOLDS with flat crenulation cleavage		D_2		D_3		D_2
Late Early Devonian intrusion of Shap (394Ma) and Skiddaw (399Ma) Granites							
(Přídolí) WINDERMERE GROUP (Mid-Caradoc)	MAIN END-CALEDONIAN PHASE: UPRIGHT FOLDS — Major and minor, with transecting cleavage, trending NE to E	F_3	D_1	Phase 3 — Related to collision	D_2	D_3	D_1
(Early Caradoc) BORROWDALE VOLCANIC GROUP (Llandeilo)	VOLCANO-TECTONIC FLEXURING AND TILTING — Open E-W folding, block faulting — INITIATION OF ENE-TRENDING LAKE DISTRICT ANTICLINE?		E-W folds large scale, no cleavage	Phase 2 — Related to subduction and closure	Not recognised in Skiddaw Group	D_2	Volcano-tectonic deformation (Branney and Soper, 1988)
VOLCANO-TECTONIC UPLIFT BEGINS? (Llanvirn) (Arenig) SKIDDAW GROUP (Tremadoc) ?	N-TRENDING FOLDS no cleavage	F_1 and F_2 (descriptions as D_1 and D_2 this volume)	N-S folds minor, no cleavage	Phase 1 — N-S folds, minor in largely unconsolidated sediments	D_1 — N-S folds, recumbent and minor, in largely unconsolidated sediments	D_1 — N-S folds (but variable), large and small scale submarine slides and slumps	D_0 — Large and small scale slumps as Webb and Cooper (1988), early small scale slumps

1972) which do not affect the overlying volcanics. Although it has been suggested (Roberts, 1971, 1973) that these north-trending folds were related to 'supposed' large-scale north-trending pre-Windermere Group (pre-Bala) structures in the Borrowdale Volcanic Group (Mitchell, 1929), it was generally believed that they represented the results of small-scale tectonic folding prior to the volcanics (but see Roberts, 1977a).

A recent resurvey of the west part of the Skiddaw Group by the Geological Survey (Webb and Cooper, 1988) has identified both major north-trending folds (amplitudes ~500 m, open to isoclinal in style, with upright to recumbent axial planes) and associated congruous minor folds (amplitudes of a few centimetres to several metres, open to isoclinal in style with straight limbs, angular to rounded hinges, and with axial planes inclined to recumbent and asymptotic to bedding). Webb and Cooper (1988) have demonstrated that the style and geometry of these early north-trending folds and associated thrusts are compatible with their generation as submarine slumps, or slide masses. They believe that this deformation was the result of major slumping of the Skiddaw Group sediments towards the central axis of a local depositional basin. The change in vergence across the Causey Pike Thrust, which trends to the SSE, suggests that this is approximately the axis of the basin. It would seem more appropriate to class these and all folds that pre-date the main cleavage as the product of soft-sediment deformation (D_0, Table 3.1).

Webb and Cooper (1988) suggested that the slumping was initiated, in Ordovician times, by listric, extensional normal faulting of the Lake District Basin, downthrowing to the north-west, away from the continent. It is possible that the Causey Pike Thrust was originally one such normal fault, reactivated as a thrust during later compressive events. Webb and Cooper further proposed that the early development of the Lake District Anticline (which has a Caledonoid trend) probably represents partial inversion of the basin due to such reversed movements. They do note, however, that volcanic doming must have been an important factor at least initially: the Borrowdale Volcanic Group–Skiddaw Group unconformity has long been related to the early development of the Lake District Anticline (Downie and Soper, 1972; Wadge, 1972).

Branney and Soper (1988) have, however, recently rejected compressive mechanisms as a means by which the Skiddaw–Borrowdale Group unconformity was generated, in view of the absence of pre-volcanic compressive structures in the Skiddaw Group. They did, however, observe that the westward overstep of the sub-Borrowdale unconformity appears to be related to the shape of the Lake District Batholith, which Firman and Lee (1986) argued, on geological grounds, was emplaced mainly in Ordovician times. Branney and Soper therefore suggested that the Skiddaw–Borrowdale unconformity was essentially volcano-tectonic in origin: the Skiddaw Group being uplifted by buoyancy effects associated with the generation of andesitic melt, by subduction (of Iapetus), and its rise through the overlying wedge of continental lithosphere. They also suggested that the slump-folding mechanism may have been responsible for the removal of part of the Skiddaw Group sequence prior to the uplift.

Pre-Windermere Group deformation phase

The Borrowdale Volcanic Group is unconformably overlain by the Windermere Group. This unconformity was first recognized by Aveline (1872) and its regional significance was established by Aveline *et al.* (1888), who demonstrated the progressive overstep of the volcanic sequence by the Coniston Limestone (Caradoc–Ashgill) south-westwards from Coniston.

The highly competent Borrowdale Volcanic Group exhibits folds of an entirely different style to those of the Skiddaw Group. There are practically no minor folds and the major folds are of large-scale open structures, with half wave-lengths of 2–7 km and limb dips varying from gentle to vertical (Figure 3.1). The geometry of these major folds is extremely difficult to define: they usually have a periclinal or monoclinal geometry, they are often so open that it is not possible to determine their axial traces, and they are frequently disrupted by faults (Branney and Soper, 1988).

The most notable major folds are the Scafell–Place Fell Syncline, easily followed for more than 30 km across the whole of the volcanic outcrop, the Ullswater, the Wrynose and the Nan Bield Anticlines, and the Ulpha Syncline (Figure 3.1).

Early workers (Green, 1920; Mitchell, 1929) suggested that the Borrowdale Volcanic Group–Windermere Group junction was marked by NNE folding of the volcanic rocks in pre-'Bala' (late Ordovician) times. Mitchell's (1929) map shows NNE-trending folds in the volcanic rocks truncated by the pre-Coniston (pre-Windermere Group) unconformity near Kentmere. This suggestion was accepted for many years, and strikes, trending NNE and N, in the Borrowdale Group (Clark, 1964; Moseley, 1964, 1972) and even in the Skiddaw Group (Roberts, 1971; Jeans, 1971) were related to a pre-Windermere Group tectonic event.

However, mapping of the Ulpha Syncline between Coniston and Dunnerdale has revealed it has an ENE, rather than a northerly, trend (Mitchell, 1940; Mitchell, 1956b). Moreover, Soper and Numan (1974) reinvestigated the Kentmere area and demonstrated that NNE-trending pre-'Bala' folds do not exist. In a theoretical reconstruction, they eliminated the presumed effects of considerable end-Caledonian deformation, believing that the Borrowdale Volcanic Group was subjected to E–W folding during the Late Ordovician. Two demonstrably open folds of this generation are the Ulpha Syncline and the Nan Bield Anticline.

A recent detailed resurvey of the Borrowdale Volcanic Group by Branney and Soper (1988) has led them to dispute Soper and Numan's (1974) findings. The age of the major folds in the Borrowdale Volcanic Group often cannot be ascertained with certainty (that is, whether they are of Caradoc or late Caledonian age) and the Ulpha Syncline remains the only major fold that is demonstrably Caradoc in age (from its relationship with the overlying Coniston Limestone) (Soper and Numan, 1974; Branney and Soper, 1988). Soper and Numan (1974) related the Ulpha Syncline and Wrynose Anticline, by means of a supposed common limb, and suggested that the Wrynose and Nan Bield Anticlines were one structure, although they could not be connected across the central Grasmere area. Branney and Soper (1988), however, now propose that the Borrowdale Volcanic Group is characterized by a large amount of block faulting and, therefore, infer that the relationship between the Wrynose and Nan Bield Anticlines is tenuous. Moreover, they calculate that the common limb between the Wrynose Anticline and the Ulpha Syncline was subhorizontal in Late Ordovician times. Thus, there appears to be little evidence for a Caradoc age for the majority of the Borrowdale Volcanic Group major folds. In fact, Branney and Soper (1988) believe that the Borrowdale Volcanic Group structure is more indicative of brittle extension than ductile compression and they suggest a volcanotectonic origin for the Borrowdale Volcanic Group–Windermere Group unconformity.

Firman and Lee (1986) have suggested that the Borrowdale Volcanics were uplifted by emplacement of the underlying concealed Lake District Batholith, this surface subsequently being covered by a Coniston Limestone Formation (Windermere Group) marine transgression. Branney and Soper (1988), however, while postulating a relationship between the batholith and the Borrowdale Volcanics, consider that the main movement at this time was a substantial, net-downward displacement to permit the preservation of some 5 km of subaerial volcanics beneath a marine sequence. They propose that the volcanotectonic faulting and tilting was associated with caldera collapse and eruption of voluminous ash flows in the upper part of the pile. The Ulpha and Scafell Synclines may well represent sags, instead of primary compressional buckles.

A volcanotectonic, rather than a compressional origin for the structures in the Borrowdale Volcanic Group is supported by the fact that structures of Caradoc age have never been reported from the Skiddaw Group. Thus, interpretations of the unconformities, both above and below the Borrowdale Volcanic Group, have, in recent times, moved away from models involving compressive tectonic events (in some cases of orogenic proportion) to volcanotectonic controls involving little, if any, tectonic folding. The only Early to early Late Palaeozoic event, therefore, which appears to have involved significant tectonic shortening deformation, is that which occurred in Early Devonian times, as a result of continental collision.

It seems, thus, inappropriate to give any previous disturbance, be it sedimentary or volcanic in origin, a 'D number', especially when this would remove the compatibility of numbering with that for the Early Devonian deformation elsewhere (for example, Wales and the Southern Uplands).

Early Devonian deformation phase

The main phase of the late-Caledonian Orogeny occurred during the Early Devonian and is characterized, in all the three main Lower Palaeozoic rock groups, by the development of steep cleavage, folding, regional metamorphism (which rarely exceeds low greenschist grade), and subsequent faulting. This can be related to the final episode in the destruction of the Iapetus Ocean with continental collision and the formation of the Old Red Sandstone (Euro-American) continent.

Table 3.1 outlines the deformation sequences of the Lake District as interpreted by various authors (Jeans, 1972; Moseley, 1972; Helm, 1970; Simpson, 1967; Soper, 1970; Roberts, 1977a; Webb, 1972; Webb and Cooper, 1988). Three phases of deformation are now generally identified as being of late-Caledonian age. The principal late-Caledonian movement (D_1) generated upright folds trending to the NE and E with associated, often strong, cleavage. These are superimposed by reclined folds and crenulation cleavage (D_2) which are widely developed in the Skiddaw Group, but only sporadically in the younger Windermere Group. This deformation is believed to have resulted from the intrusion of the Lake District batholith (Roberts, 1977a). Finally, minor N–S flexures and fracture cleavage (D_3) developed during axial shortening (Roberts, 1977a), especially in the Skiddaw Group.

Deformation characteristics

The following sections discuss the characteristics and implications of the late-Caledonian deformation (folding, cleavage, faulting) in the major rock groups of the Lake District. Differences in the style and scale of the deformation are related to previous strain history (for instance, the Skiddaw, Eycott, and Borrowdale Groups had already been variably deformed, whereas the younger Windermere Group had not) and, more importantly, bedding anisotropy (particularly layer thickness and competence contrasts) (Moseley, 1972).

Folds

Both major and minor F_1 folds are present in the Skiddaw Group. Major F_1 fold axial traces show a predominant NE or ENE trend and are associated with an approximately axial-planar cleavage. Major folds have not been traced for large distances, except for the 'Lake District Anticline', which has its axial trace within the Skiddaw Group, between the opposing dips of the Eycott and Borrowdale Volcanic Groups – see Figure 3.2, Line A–B. Webb and Cooper (1988), in their detailed investigation of the west part of the Skiddaw Group, could not demonstrate interference between major F_0 and F_1 folds, although interference on a minor scale is commonly seen and on an intermediate scale at Hassness. Webb and Cooper (1988) suggested that at least some F_1 folds could represent modified F_0

folds: where the slump folds are parallel to the main Caledonian trend, they are tightened and can develop an approximate axial-planar cleavage, but where they are of different trend, the cleavage is oblique to fold axes. This clear superimposition is exemplified at both Hassness and Gasgale Crags, although some ambiguity remains at Buttermere and Warnscale Bottom. Slump folds in a granite aureole with little modification by D_1 are preserved in the River Caldew.

Minor F_1 folds in the Skiddaw Group are best developed in psammite–pelite layers (for example, the transition beds between the Loweswater Flag Formation and the Mosser Slate Formation). These can be seen at a number of the sites described below, but are best seen at Raven Crags. They tend to be upright folds with amplitudes of a few metres or less. Since they are generally parasitic to major folds of the same generation, they are often asymmetrical. Mudrocks are usually more complexly folded and it is often difficult to distinguish the F_1 structures. Where they can be observed (for instance, Warnscale Bottom) they are upright, tight to open with gentle plunges to the north-east and south-west and disharmonic effects are common (Soper and Moseley, 1978). The F_1 folds in massive arenaceous units, such as those towards the base of the Loweswater Flag Formation, are open folds which frequently approach a true 'parallel' style and have interlimb angles often greater than 90°. The later phase of folding (F_2) is of an open recumbent nature, generally with a near-horizontal crenulation cleavage, and is restricted to beds which are close to vertical, as at sites at Buttermere and Raven Crags.

The competent Borrowdale Volcanic Group is characterized by fold structures very different from those in the Skiddaw Group: large open folds are associated with brittle fracture on all scales. The most persistent of these are the Scafell, Haweswater, and Ulpha Synclines and the Wrynose and Nan Bield Anticlines – see Figure 3.2, Line C–D. Their long, linear nature is very much like that of the main folds crossing the Windermere Group and, in view of Branney and Soper's (1988) recent ideas on the Borrowdale Group (see above), it is most likely that their main period of formation was by compression during the late-Caledonian Orogeny, rather than the more restricted folding associated with the volcano-tectonic doming. Only the Ulpha Syncline (see Limestone Haws) can positively be attributed to

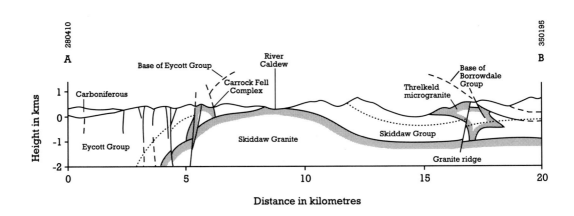

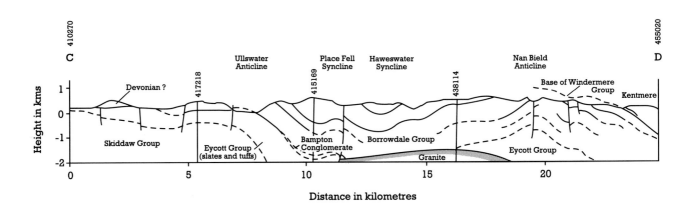

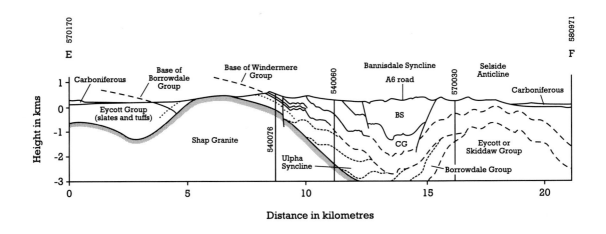

BS   Bannisdale Slates

CG   Coniston Grits

Figure 3.2   Cross-sections along lines shown in Figure 3.1 (modified from the work of N. J. Soper in Johnson *et al.*, 1979).

this latter event, both its limbs being cross-cut by the Windermere Group (Branney and Soper, 1988).

Medium-scale folds in the Borrowdale Volcanic Group (for example, the Yoke Folds near Kentmere) were probably initiated during end-Caledonian movements (Soper and Numan, 1974). Minor structures are not common in the relatively competent Borrowdale Group, although small-scale folds, roughly congruous with the cleavage, are developed occasionally in the more thinly bedded parts of the volcanic succession. However, it is generally not possible to unambiguously ascribe them to the main late-Caledonian D_1 event (Soper and Moseley, 1978).

Since the Windermere Group and the Skiddaw Group are made up of very similar lithologies, the F_1 fold styles in these two are very similar. Thus, the comments already made regarding development of F_1 folds in the Skiddaw Group are generally pertinent to the Windermere Group, except that the Skiddaw Group had already been subjected to the slump deformation (D_0), prior to the D_1 late-Caledonian deformation, and can locally exhibit a very complex polyphase structure, as discussed above. The Windermere Group shows a much less-complex structure and is rather more simple to interpret.

For four or five kilometres south-east of the Windermere Group unconformity the rocks dip steeply to the south-east, as at Limestone Haws. Lack of small-scale folding may have been influenced by the large amount of massive grey-wacke in this part of the sequence, and by the underlying volcanics. South of this area, the Bannisdale Slate Formation, in particular, is strongly folded in the form of synclinoria and anticlinoria.

Major F_1 folds (for example, the Bannisdale Syncline and the Selside Anticline; Figure 3.2, Line E–F) can only be determined by regional mapping and their axial traces are usually determined from the asymmetry of the minor folds, as can be demonstrated at Shap Fell and Tebay. Minor folds are common in the Bannisdale Slate Formation, in the pelite–psammite layers transitional between the Bannisdale Slate Formation and the Coniston Grit Formation, but occur less frequently within the more massive and competent Coniston Grit such as at Tebay. Minor folds have half wavelengths which vary from a few metres to about 200 m; they are periclinal in form, dying out as conical structures (Webb and Lawrence, 1986; Lawrence *et*

al., 1986). The plunges of the folds are also quite variable across the Windermere Group. Near Coniston, in the west, fold plunge is about 30° to the north-east (Soper and Moseley, 1978; Moseley, 1986): while in the east of the Lake District there is a low plunge of less than 5° to the ENE which has been attributed to post-Carboniferous tilt (Moseley, 1968, 1972). At Helwith Bridge plunges are 15–20° ESE.

Cleavage

Main Caledonian cleavage (S_1) affects all the major rock groups in the Lake District. The general trend is NE–SW, but in detail is arcuate (Figure 3.3). The strike swings from N–S in the Grange-over-Sands area in the south, ~050° in the south-west (near the Duddon estuary), through 080° in the vicinity of Kendal and 090° in the northern Howgill Fells, to 105° in the Ribblesdale inliers (Soper *et al.*, 1987). This swing can be demonstrated at Limestone Haws, Shap Fell, Tebay, and Helwith Bridge. 3.3

Within the Skiddaw Group, cleavage is strong in pelites and weak in psammites. Spaced cleavage predominates and it is unusual to find a truly penetrative fabric. The main cleavage usually dips at a high angle and is frequently parallel to bedding, so that axial-planar cleavage is not common, and only seen locally on fold crests. It is not clear how extensive is the true 'bedding-cleavage' recorded by Roberts (1977a). A later near-horizontal crenulation cleavage can often be seen associated with the open (F_2) recumbent folds (Buttermere, Gasgale Crags, and Raven Crags).

Cleavage in the Borrowdale Volcanic Group is also related to lithology and is strong in the fine-grained volcaniclastic tuffs (Hollows Farm and Jumb Quarry), but weak in the more massive lavas and sills (Warnscale Bottom and Limestone Haws). There are also several zones of strong cleavage (high strain) and poor cleavage (low strain), which Firman and Lee (1986) have suggested to be related to the roof of the Lake District batholith. Cleavage is poorly developed where the batholith is near the surface and strong where it is deeply buried or absent. For example, there is a high-strain zone with strong cleavage running through Honister, where there are important slate quarries, and yet cleavage is almost non-existent on the adjacent High Stile range.

In the Windermere Group, cleavage varies from

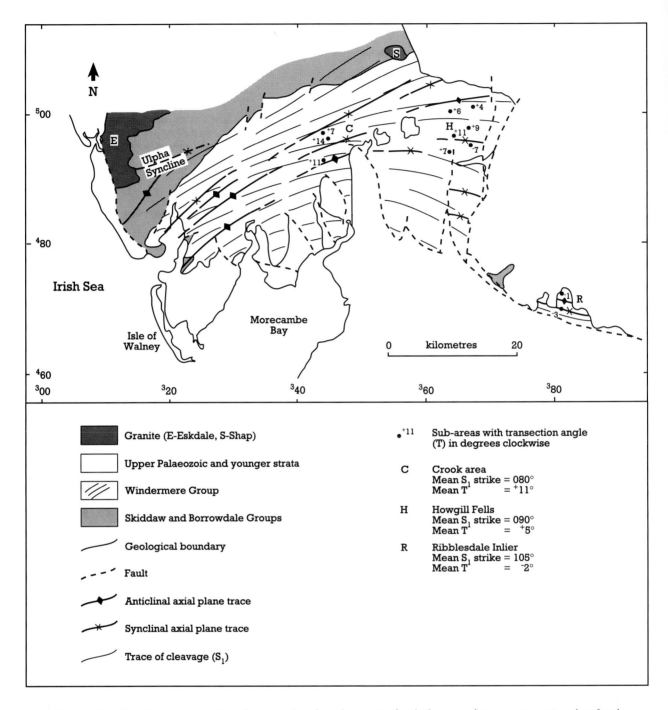

Figure 3.3 The cleavage arc in the Silurian rocks of north-west England, showing cleavage transection data for the Crook area, eastern Howgill Fells and Ribblesdale Inlier (after Soper *et al.*, 1987).

a strong fracture cleavage in pelites, to none at all in massive greywackes. Refraction can usually be seen in graded greywacke units, and this is seen especially well at Shap Fell. Interestingly, the end-Caledonian cleavage is generally not axial-planar to associated folds. The cleavage strike transects the axial planes of the folds, usually a few (~5) degrees clockwise (Moseley, 1968, 1972; Lawrence *et al.*, 1986; Shap Fell). This clockwise transection is believed to result from sinistrally oblique compression (transpression, Harland, 1971). Soper *et al.* (1987) have examined this transection throughout north-west England and demonstrated that the angle diminishes towards the east of the

Lake District, to become anticlockwise at Helwith Bridge. These authors also observed that arcuate structures of the Lake District appear to have been moulded around the northern flank of the Midlands Massif. They believe the latter acted as a rigid indenter, about which the Lower Palaeozoic rocks were deformed as Avalonia was accreted northward on to the margin of Laurentia (Soper *et al.*, 1987, Figure 2).

Studies to quantify the strain associated with slaty cleavage formation, using accretionary lapilli in tuffs from the Borrowdale Volcanic Group, were initiated by Green (1920). More recent investigations have been carried out by Oertel (1970), Helm and Siddans (1971), and Bell (1975, 1981, 1985) at Jumb Quarry. There has been controversy about this work, the debate centring on the origin of slaty cleavage (Soper and Moseley, 1978) and whether it relates to the total (finite) strain to which a rock has been subjected in its history, or reflects a particular component of the total strain. It would appear that, in the case of sedimentary rocks which underwent volume reduction during compaction before their tectonic deformation, the latter must be true (Soper and Moseley, 1978). Taking the above into account, in his modelling of high-strain zones at Jumb Quarry, Bell (1975, 1981) suggested a maximum compaction strain of ~66% normal to the bedding, prior to tectonic strain (close to plane strain) which resulted in shortening of ~50–70% across cleavage.

Faults

Important fault movements occurred during late-Caledonian deformation events. In some areas these movements probably reactivated older faults, but this is difficult to prove. It is evident that the late-Caledonian folding and faulting were to some extent synchronous (Moseley, 1968). Several types of faulting have been identified. Thrust faults are well-developed close to the lower and upper junctions of the Borrowdale Volcanic Group (see below, Warnscale Bottom and Hollows Farm). Important wrench faults have displacements from a few metres to 1–2 km (see Limestone Haws), and composite wrench-thrust faults, first mapped in detail by Norman (1961), are basically north-trending, sinistral wrench faults which bend into thrusts. They trend to the north-east, are inclined at ~45° to the south-east, and then bend back into wrench faults forming dog-leg outcrops (Moseley, 1972).

An important fact to emerge from the study of late-Caledonian faults in this area (as well as the rest of the British Caledonides) is that displacement on NE–SW shear zones (Hutton, 1982) and strike-slip zones (Watson, 1984) was sinistral, not dextral. This has important implications for plate tectonic models for the closing of Iapetus – see Chapter 1.

Timing of late-Caledonian movements

In previous literature, late-Caledonian events are frequently referred to as being of 'end-Silurian' age. This dating arose from the observations that the main Caledonian deformation affected all the Lower Palaeozoic rocks of the Lake District, up to the youngest Silurian strata (the Scout Hill Flags, deposited in Pridoli times), but did not affect the molasse-type Mell Fell Conglomerate of uncertain Devonian age (Capewell, 1955; Wadge, 1978b). Recently however, Soper *et al.* (1987) have pulled together evidence which strongly indicates that the main Caledonian deformation occurred in Early Devonian times. For example, the Shap Granite post-dates the main cleavage (S_1); contact minerals in the aureole having grown across cleavage planes, yet the cleavage itself is deflected around the granite (Boulter and Soper, 1973). The Skiddaw Granite also post-dates the main cleavage; andalusite in the contact aureole clearly having overgrown the main cleavage fabric. In places, however, the latter shows weak contact strain around porphyroblasts, which implies that it began to grow during the waning stages of compression. Soper *et al.* (1987) believed this evidence to indicate that the Shap and Skiddaw Granites were emplaced during a period of stress relaxation, immediately following the main compressive phase of late-Caledonian deformation (see also Soper, 1986). As the isotopic age of the intrusions allows assignment to the Early Devonian Period (Shap Rb–Sr age = 394 ± 3 Ma (Wadge *et al.*, 1978); Skiddaw K–Ar biotite age = 392 ± 4 Ma (Shepherd *et al.*, 1976), Rb–Sr age = 399 ± 8 Ma (Rundle, 1981)), Soper *et al.* (1987) suggest that the main deformation in the Lake District was also, most probably, Early Devonian in age. Further evidence from Wales, considered by Soper *et al.* (1987), McKerrow (1988) and Soper (1988), suggests that the deformation was Emsian in age, equivalent to the Acadian Orogeny of the Canadian Appalachians.

Tectonic models

The plate tectonic model of Dewey (1969), and most of its subsequent variations, envisage the Early Palaeozoic evolution of the Lake District as taking place on the north-western margin of the Avalonian continent (see Figure 1.2). The argument for the position of the Lake District to the south of the suture, now positioned beneath the Solway Firth, and over a south-easterly dipping subduction zone, depended partly on the lithological character of the Skiddaw Group and its 'European' fauna (Fortey, 1989), but mostly on the presence of the volcanic arc represented by the Borrowdale Volcanic Group. Support for the latter aspect of the model came from the observation of Fitton and Hughes (1970) that a southerly directed subduction zone could be inferred from the change from the tholeiitic volcanics of the Llanvirn Eycott Group to the calc-alkaline character of the Caradoc Borrowdale Volcanic Group.

The only specific structural characteristics that have been used to support the model are the intra-Ordovician tectonic shortenings represented by the two deformation episodes that pre-date the Borrowdale Volcanic Group and the Windermere Group, respectively. Otherwise, the general view has been (Moseley, 1977) that, after initial gradual closure of the ocean in the Late Ordovician, there was a final collision in the Late Silurian to Early Devonian, resulting in the D_1 folding and cleavage.

The evidence for pre-D_1 shortening has now been reinterpreted, following the general rejection of Simpson's (1967) proposal for major tectonism prior to deposition of the Borrowdale Volcanic Group, as discussed above. There has been the recognition, firstly, that most pre-D_1 folding in the Skiddaw Group is of soft-sediment origin (Webb and Cooper, 1988), and, secondly, that the pre-Windermere Group folding affecting the volcanics can be attributed to caldera collapse and block-tilting (Branney and Soper, 1988). Not only do these authors reject the evidence for tectonic shortening, but they emphasize the probable importance of extensional deformation in the development of this folding.

Apart from the D_1 shortening witnessed by folding and cleavage, there is also evidence of thrusting, beneath the Borrowdale Volcanic Group, within the Stockdale Shale Formation and locally elsewhere, where there are competence contrasts. The scale of the movement is not clear, and neither is it clear whether any of it could be of pre-D_1 age. Moseley (1972) associates some of the thrusting with post-D_1 faulting, that is, post-Emsian but pre-Carboniferous faulting.

The plate tectonic context of the Lake District still rests firmly on its general setting in the British Caledonides (see 'Introduction', Chapter 1). However, reassessment of its precise role will undoubtedly take place as a result of the recent work, quoted above, and work in progress. Particular studies will be significant in this respect, namely, the evolution of the Skiddaw Group Basin(s) and its relation to volcanicity and extensional faulting; the relation of the sub-Borrowdale Volcanic Group unconformity to soft-sediment deformation, volcanic doming, the underlying batholith, and the initiation of the Lake District Anticline; the development of the volcanic arc, the polarity of which has recently been queried by Branney and Soper (1988). Further development can be expected from the recent discussion by Soper *et al.* (1987) of the arcuate pattern of D_1 deformation, the change in the cleavage/fold transection angle, and the relationship of these features to the geometry and motion of the Avalonian continent.

BUTTERMERE VILLAGE (NY 170176)
F. Moseley

Highlights

Several exposures, adjacent to Buttermere Village, reveal all the small-scale structures generally observed within the Skiddaw Group. At least three phases of deformation are recorded in the rocks here: F_0 folds, generated by slumping, are refolded by F_1 folds of the main D_1 late-Caledonian deformation; crenulation cleavage of D_2 age is also present.

Introduction

The easily accessible exposures around Buttermere Village illustrate well, many of the structural problems of the Skiddaw Slates. In the past, attempts have been made to explain most of the structures hereabouts in tectonic terms (Simpson, 1967; Moseley, 1972; Webb, 1972; Soper and Moseley, 1978). Although obvious sedimentary slumps, typical of many turbidite systems, have been recognized for many years (see Introduction, Chapter 1) the slates have seemed to have three distinctive phases of folding (all presumed to be

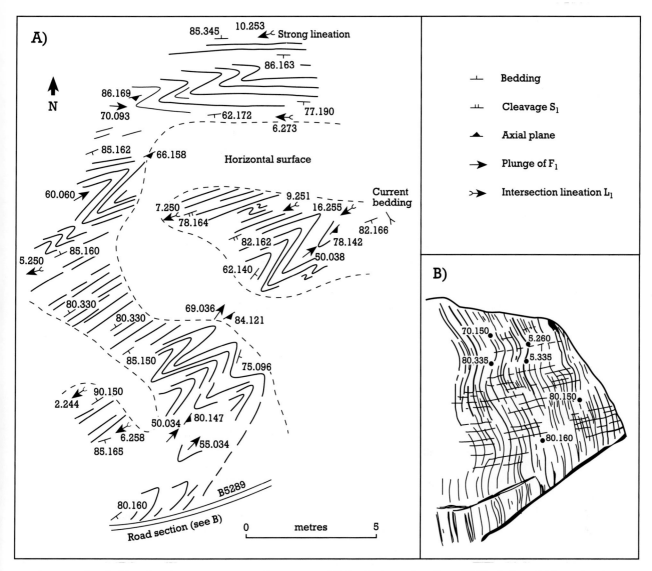

Figure 3.4 Skiddaw Group exposures, near Buttermere. (A) is a horizontal surface. (B) Vertical roadside section in (A) looking towards 060°. Three fold phases are represented in these exposures. The steep plunge of the folds represents the dip of a limb fold, initiated during F_0. The D_1 phase is represented by the tight ENE–WSW folds and related cleavage, and D_2 by open recumbent folds and crenulation cleavage which can only be viewed on vertical surfaces, where the other two phases cannot be seen (after Moseley, 1981, and notes by D. Aldiss B.Sc. thesis, Birmingham University, 1974).

tectonic), which have been previously labelled 'F_1', 'F_2', and 'F_3' (see Table 3.1). Folds identified as 'F_1' are small-scale, complex and have no associated cleavage. They have a variable trend, often N–S, and until recently they were believed to represent a pre-volcanic phase of minor folding (Jeans, 1972; Moseley, 1972; Webb, 1972). The F_2 folds are similar small-scale structures which resemble the F_1 structures, except that they have a generally NE to ENE trend and often have axial-planar cleavage. Most authors have attributed them to the main end-Silurian phase of the Caledonian Orogeny.

The F_3 folds are open recumbent structures with a subhorizontal crenulation cleavage which have also been thought to belong to the main Caledonian deformation.

Description

This site shows the muddy siltstones of the Buttermere Formation of the Skiddaw Group. At Buttermere Quarry (NY 17331727), the strata are inclined about 70°SE, with load casts near the top of the quarry showing the way-up. S_1 cleavage and bedding are parallel, but there are near horizontal F_2 crenulations.

On Long How (NY 17251730), outcrops of pelites with silty layers expose steeply plunging folds. These folds, once thought to be tectonic (Moseley, 1972) are now thought to be slumps refolded by the main D_1 movements.

Alongside Millbeck (NY 17001717), an F_1, vertically plunging fold is seen to be refolded by an F_2 recumbent fold (Moseley, 1972). Intrafolial folds are seen along one of the F_1 fold limbs.

The fourth locality (NY 17651703; Figure 3.4) is at Buttermere Church, where the gentle surface of a roche moutonée displays steeply plunging folds. The steep plunge is thought to represent a steeply dipping limb of a slump fold (F_0), whereas the folds themselves are largely the product of the D_1 late-Caledonian deformation. These tight folds are related to an S_1 cleavage. The vertical surface alongside the road reveals open recumbent folds (F_2) with a weak crenulation cleavage (Moseley, 1981).

Interpretation

Recent changes in interpretation by Webb and Cooper (1988; and see below – Hassness and Gasgale Crags) are that the F_1 and even some F_2 small-scale folds originated as slumps, those with the main Caledonoid north-easterly trend having been subsequently tightened and developing a cleavage during the main orogenic phase (now considered to be Early Devonian). The difficulty in the field arises from the lack of unambiguous criteria for assessing the origin of the F_1 and F_2 folds. In this description, the F_1 folds, considered by Webb and Cooper (1988) to be of slump origin, are designated F_0, while the F_2 folds, which seem to be coeval with cleavage of late-Caledonian age, are designated F_1. Later folds associated with a flat crenulation cleavage are consequently designated F_2.

Although interference between F_0 and F_1 folds in these localities is limited, it is clear that three sets of structures are represented. The F_0 and F_1 folds are both tight and steeply plunging, but F_0 has S_1 cleavage superimposed. The steep plunge of F_1 can be attributed to a steep dip, produced by large-scale, slump folds (F_0). The interpretation of the F_0 minor structures and steep pre-F_1 dips as indications of sedimentary slump processes rests largely on the arguments of Webb and Cooper,

(see above). These authors show that the folds are related to major folds and olistostromes which have variable trend and vergence, but which pre-date the D_1 folds and cleavage.

The Buttermere outcrops show, particularly clearly, that the third set of structures are superimposed on the earlier two. The D_2 affects steeply dipping surfaces to give open and re-cumbent F_2 folds and a related flat S_2 crenulation, although attitudes depend on the dip of the affected surface.

The D_1 and D_2 structures are considered (Webb and Cooper, 1988) to be a product of the late-Caledonian deformation. It now appears that there are no significant tectonic folds that pre-date the Borrowdale Volcanic Group.

Conclusions

The Buttermere site is important in that, in adjacent outcrops, all three sets of structures that are common to much of the Skiddaw Group can be demonstrated. The first of these (isoclinal folds) were produced by slumping, that is, move-ment and deformation of masses of sediment either contemporaneous with, or relatively soon after, their deposition on a sloping sea-bed, which means that these folds are of Ordovician age. The second generation of folds (main phase, tight steeply plunging folds) were formed during the main Caledonian mountain-building 'storm', during the early Devonian, at about the time of the closure of the Iapetus Ocean through the convergence of the landmasses to its north and south. The third-generation (open recumbent) folds are taken to be late Caledonian. All three categories are important in the context of the Caledonian evolution of the Lake District.

HASSNESS AND GOAT CRAGS (NY 189163)
B. C. Webb

Highlights

This well-exposed section, showing some 350 m of the Buttermere Formation of the Skiddaw Group, demonstrates the distinction between the disrup-tion and folding produced by slump or gravity-slide movements during Llanvirn times and the superimposed Early Devonian tectonic deformation.

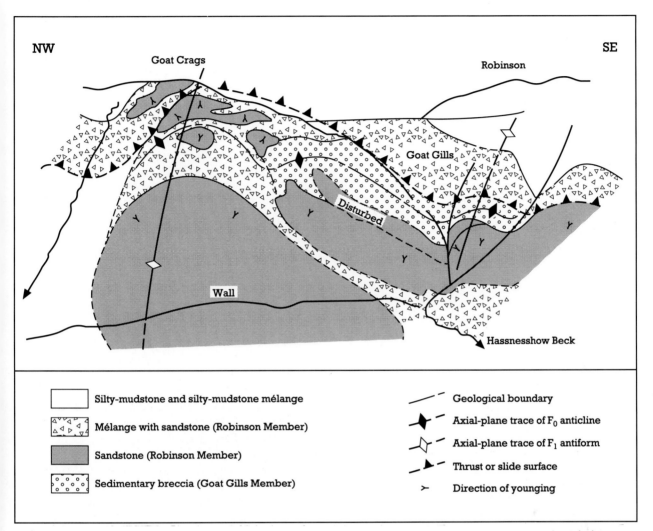

Figure 3.5 Hassness and Goat Crags. Sketch view from the south-west, showing outcrop pattern and axial-plane traces. Length of foreground is approximately 500 m.

Introduction

The Skiddaw Group (Lower Ordovician) rocks exposed here are disposed on a major anticline, overturned broadly westwards. The uninverted limb of this fold has suffered severe extension, leading to disruption of the beds and the formation of mélange lithologies.

The site lies within ground described by Rose (1954), Jackson (1961, 1962, 1978), Simpson (1967) and Webb (1972, 1975). Because of the complex structure and lack of chronostratigraphical control, the results of these earlier workers were controversial and problematic. The recent resurveying of the western part of the main Skiddaw Group inlier by the British Geological Survey has resolved the structural problems and clarified the stratigraphy (Webb and Cooper, 1988).

It should be noted that, in the description below, the gravity-driven folds labelled F_0 and the tectonic deformation labelled D_1 are differently assigned by Webb and Cooper (1988) – see Table 3.1. This merely results from an attempt, in this volume, to reserve the notation D_1 for the Early Devonian deformation phase, which was broadly contemporaneous throughout the non-metamorphic Caledonides.

Description

The strata at the site belong to the Buttermere Formation of the Skiddaw Group comprising, in ascending stratigraphical order:

1. The Goat Gills Member, a marine breccia which has yielded a Tremadoc microflora.

2. The Robinson Member, a sequence of inter-bedded turbidite sandstone and siltstone.
3. Undifferentiated silty mudstone, largely of Late Arenig age.

Two generations of folds are clearly visible in the rocks cropping out over the site area, designated F_1 and F_2 by Simpson (1967), but are here designated F_0 and F_1, in accordance with the regional deformation sequence (F_1 and F_3 in Webb and Cooper, 1988). Both sets of folds are displayed in the Robinson Member, where bedding is clearly visible and the 'way-up' of the strata can be easily ascertained from sedimentary structures (Figure 3.5). The Robinson Member exhibits a strong ductility contrast with the surrounding silty mudstone, and this facilitates the study of deformation structures associated with slumping and with the generation of mélange structures (Figures 3.5 and 3.6).

Both major and minor F_0 folds are present. The trace of a major F_0 anticline descends from near the top of Goat Crags, south-eastwards through Goat Gills (Figure 3.5). This anticline is overturned, broadly westwards, so that the beds cropping out over most of the site dip eastwards and are inverted. Minor F_0 folds with amplitudes of a few metres, or less, are common. They are intrafolial periclines with curvilinear hinges, which plunge south-eastwards and are congruous with the major fold. On the inverted limb of the major F_0 fold, turbidite sandstones of the Robinson Member crop out extensively (Figure 3.6). The sandstones are hardly disrupted except close to their junction with the stratigraphically underlying Goat Gills Member, which crops out in the core of the major fold. Near to this junction, sandstone beds are irregularly and disharmonically folded and sheared and only very locally does the junction appear to be undisturbed. On the more gently dipping, uninverted limb of the major fold, the Robinson Member is highly disrupted and forms a mélange of sandstone rafts and boudins suspended in a silty-mudstone matrix (Figure 3.7). Sandstone 'rafts' are well exposed near the summit of Goat

Crags, where they range, in length and thickness, from a little over 1 m to several tens of metres.

The F_1 folds occur, most commonly, as minor structures with amplitudes of only 1 or 2 m. Larger F_1 folds, a few tens of metres in amplitude, affect the Robinson Member on the inverted F_1 major fold limb towards the west end of the site and in the buttress between Goat Gill and Hassnesshow Beck (Figure 3.5). The main cleavage (S_1) is well developed in the finer-grained lithologies. It dips steeply to the SSE, axial planar to the F_1 folds. Interference between F_0 and F_1 folds is well displayed at Hassness. The F_0 fold hinges, with gentle plunges, are visible in the small cliffs overlooking the lake, and F_1 folds, with axial-planar S_1 cleavage, can be seen plunging steeply in the glacially scoured rock surface above the cliffs.

Other minor structures in the Skiddaw Group are poorly represented in the site area. These include late, sideways-closing minor folds (F_2) with an associated, axial-planar, crenulation cleavage (S_2) which commonly affect S_1 or bedding, where this is steeply inclined. Sporadic, steeply inclined, NNW-trending joints are the local expression of north- or north-west-trending joints and cleavage, which are better developed elsewhere.

Rose (1954) recognized 'severe overfolding and thrusting' in the Goat Crags area. Using 'way-up' evidence, from the sedimentary structures in the turbidite sandstone, he correlated the Robinson Member with the lithologically similar Loweswater Formation (Early Arenig) of the fells further north. Jackson (1961, 1962) initially agreed with this correlation, having obtained graptolites from near the summit of Robinson (NY 202168) which, elsewhere, occur at the base of the Loweswater Formation. Rose correlated the silty mudstone with the Kirkstile Formation, which overlies the Loweswater Formation. Later, however, Jackson changed his correlation (Jackson, 1978). Ignoring the 'way-up' evidence but taking into account graptolites reported from near Buttermere Village (Simpson, 1967), he reassigned the silty mudstone to the Hope Beck Formation which underlies the Loweswater Formation. Simpson, too, largely ignored the 'way-up' evidence placing the silty mudstone, which he named the Buttermere Slates, stratigraphically below the Robinson Member. He did not correlate the Robinson Member with the Loweswater Formation, but considered it to be older, naming it the Buttermere Flags.

Figure 3.6 Goat Crag, Buttermere. Slump-generated minor folds on the inverted limb of a major slump fold in the Skiddaw Group (hammer for scale, middle right). (Photo: reproduced by permission of the Director, British Geological Survey: NERC copyright reserved, D 3843.)

Interpretation

Simpson (1967) recognized the polyphase nature of Skiddaw Group deformation, identifying the main, ENE-trending cleavage and folds (the present S_1 and F_1) as the earliest structures. He considered that these were originally inclined towards the SSE, but were reoriented by the later sideways-closing folds (the present F_2). He considered that both of these deformations pre-dated the Borrowdale Volcanic Group, which unconformably overlies the Skiddaw Group. Since the ENE-trending, main cleavage and associated folds at Goat Crags affect previously inverted strata, they cannot be the earliest folds. Simpson referred them to a later, post-volcanic deformation phase even though, elsewhere, he considered a north-westerly trend to be characteristic of this phase. Simpson's deformation sequence and the nature of the junction with the overlying volcanic rocks were the subjects of controversy during the late 1960s and early 1970s (see Soper and Moseley, 1978; and the Introduction to this chapter). No major structures relatable to Simpson's second and third deformation phases could be substantiated, and his first two deformations were shown to be Early Devonian in age and thus not related to pre-Borrowdale unconformity. Early, north-trending, pre-cleavage folds (the present F_0) were, however, discovered at various localities in the Skiddaw Group.

Webb (1972, 1975) mapped the major and minor pre-cleavage folds in Goat Crags and described, in detail, the interference between minor, pre-cleavage folds and later, cleavage-related folds at Hassness. Since the later folds and cleavage trend ENE, parallel to the main cleavage elsewhere in the Lake District, he considered these to be the Early Devonian structures.

The recent remapping was done by the Geological Survey, when a more detailed palaeontological investigation was undertaken. This demonstrated abrupt changes in the age of contiguous strata in the Buttermere area, indicating severe disruption of the normal stratigraphical sequence. Disruption of the Robinson Member

had been noted by Webb (1975) but, at that time, major submarine gravity slide deposits (olistostromes) had yet to be described in detail and he considered the deformation to be 'orogenic'. The importance of slump folding associated with major, gravity sliding of the Skiddaw Group was first clearly demonstrated by Webb and Cooper (1988). They showed not only that the north-trending folds were slump-generated, but also that many of the ENE-trending folds were early slump structures modified by later, Early Devonian deformation. They proposed the current stratigraphy, defining the Buttermere Formation as an olistostrome, or submarine slump mass. Evidence from near Causey Pike (NY 218209), further north, indicates that the olistostrome was emplaced during the Early Llanvirn. The geometry of the slump folds within it indicate that it slid westwards. This section is situated to the south-east of the Crummock Water–Causey Pike Line and the sense of overturning here is contrary to the south-easterly overturning observed north-west of that structural line (for example, at Gasgale Crags), suggesting that the line represents the axis of a local Ordovician depositional basin.

Conclusions

The fellside at Goat Crags affords an excellent section through a major olistostrome or submarine slump mass. This was formed by massive lateral movement of material on a sloping sea-bed during the early Ordovician Period, in Llanvirn times. Deposits of this type, on this scale, have not been recorded elsewhere in Britain. Primary minor structures, developed during the emplacement of the slump mass, are clearly displayed. These folds are referred to as D_0 folds in the classification of folds used in this volume. A degree of stratigraphical control within the slump mass is provided by the sandstone of the Robinson Member, whose sediments provide way-up evidence. In contrast to the south-easterly movements recorded at Gasgale Crags (see below), the slump folding here was directed to the west, towards the centre of the Ordovician marine depositional basin. The slump structures are clearly overprinted, that is, refolded, by others formed during the Early Devonian Caledonian deformation (D_1). It was here that these two generations of folding were first recognized and explained. They provide an important key to the understanding of the geological structure and history of the Lake District.

Figure 3.7 Goat Crag, Buttermere. The sandstone lens is part of a slump-generated mélange which has been folded by minor D_1 folds with a poorly developed axial-planar cleavage. (Photo: reproduced by permission of the Director, British Geological Survey: NERC copyright reserved, D 3849.)

GASGALE CRAGS AND WHITESIDE (NY 170220)

A. H. Cooper

Highlights

Slump and gravity-slide structures, which formed through south-eastwards movement, towards the centre of the early Ordovician basin, are here refolded by late-Caledonian tectonic structures. The Gasgale Thrust probably post-dates formation of the Crummock Water Aureole, dated to *c.* 400 Ma, making it a late-Caledonian structure.

Introduction

Gasgale Crags and Whiteside afford excellent exposures which illustrate the characteristic structure of the Skiddaw Group (Early Ordovician) in the north-west of the Skiddaw Inlier. Some structures, previously interpreted as of tectonic origin, can be shown to be slump folds in the turbidites. Late-Caledonian folding, cleavage and thrusting are superimposed on these early structures.

The Skiddaw Inlier was originally surveyed by Ward (1876). The stratigraphy of the Gasgale–Whiteside area was first elucidated by Rose (1954), who recognized the Loweswater Flags overlain by the Mosser–Kirkstile Slates of Early Ordovician age. He also described the major tectonic style and the metamorphism of the Crummock Water Aureole. Jackson (1961, 1978) reviewed the stratigraphy of the Skiddaw Group, and recognized the Hope Beck Slates below the Loweswater Flags. An alternative stratigraphy was suggested by Simpson (1967); although his scheme has proved untenable, he was the first to identify polyphase deformation in the Skiddaw Group. His deformation sequence has since proved incorrect. Moseley (1972) gives details of the Whiteside site, in an overview of the polyphase deformation in the Skiddaw Group; Jeans (1974) also gives local details.

Description

The current interpretation of the stratigraphy and structure is that the bulk of Whiteside is composed of Loweswater Formation greywackes, about 900 m thick. These are overlain by the siltstones of the Kirkstile Formation, in excess of 1000 m thick, which is well exposed on Gasgale Crags (Figure 3.8). The Loweswater Formation is of Early Arenig age and the Kirkstile Formation for the most part of Late Arenig age.

The top of the Loweswater Formation and most of the Kirkstile Formation are affected by synsedimentary and early post-sedimentary slump folds. The sequence is refolded by F_1 and F_2 tectonic folds (F_3–F_4 of Webb and Cooper, 1988). The tectonic folds (F_1), trend to the ENE and NE and have an associated cleavage (S_1); local developments of NW-trending cleavage (S_1?) and low-angled cleavage (S_2) and sideways-closing folds (F_2) also occur.

It will be seen (Table 3.1) that Webb and Cooper (1988, Table 1) propose a different 'D' (deformation) terminology from that used here. In this volume, D_1 (F_1, etc.) is assigned to the broadly contemporary Early Devonian deformation in the Caledonides, and D_0 (F_0) is used for gravity-driven deformation. The significance of the major lobate folds of the Darling Fell area, overturned to the south-east (designated F_1 by Webb and Cooper, 1988) has recently been reassessed. It is probable that they are in fact of F_1 and not F_0 origin, but they are broadly coaxial with the definite slump folds.

The Gasgale Crags are cut by a southerly directed thrust fault. The district is crossed by the elongate EW-trending Crummock Water Aureole (Cooper *et al.*, 1988), the northern margin of which is exposed at the west end of Gasgale Gill.

Loweswater Formation

On Whiteside, the Loweswater Formation is exposed, dipping steeply southwards. Its basal contact with the Hope Beck Formation occurs on Dodd (NY 169232) 1 km north of Whiteside. The bulk of Whiteside End is composed of medium- to thick-bedded greywacke beds, decreasing in thickness upwards (NY 16602169). The formation is dominantly quartz-rich greywacke, and throughout exhibits well-developed sedimentary structures (Bouma, 1962) indicative of deposition from distal turbidity currents. Palaeocurrent indicators (flute and groove casts) show a southerly source (Jackson, 1961). The Loweswater Formation has an Early Arenig age, mainly within the *Didymograptus deflexus* and *D. nitidus* biozones.

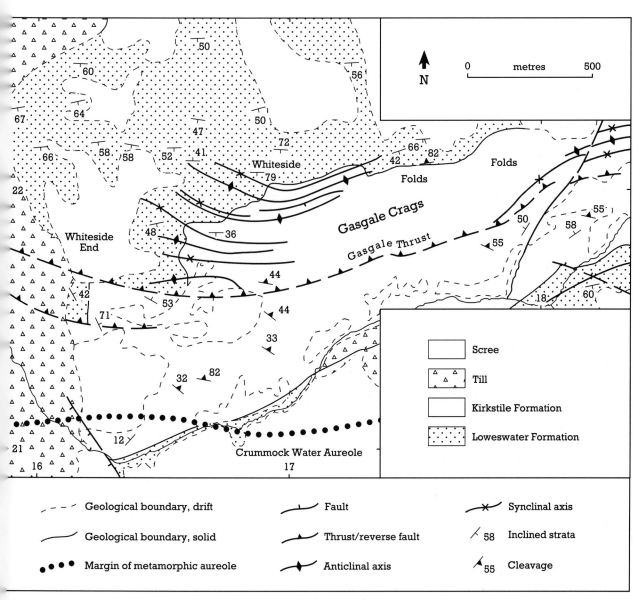

Figure 3.8 Geological map of the Gasgale Crags and Whiteside area, based on Geological Survey map (NY 12 SE) surveyed by P. M. Allen, A. H. Cooper and B. C. Webb (see also Moseley, 1990, Figure 20).

Kirkstile Formation

The Kirkstile Formation, in excess of 1000 m thick, is dominantly siltstone with mudstone beds. Near its base, subordinate thin beds of quartz-rich greywacke (Bouma C units) also occur. The proposed type section for the base of the formation, and the rapid interbedded transition from the Loweswater Formation, is exposed on Whiteside End (NY 16602169). The formation is poorly fossiliferous, of Late Arenig age, and yields evidence for the *Isograptus gibberulus* and *D. hirundo* biozones.

Slump folds

The upper part of the Loweswater Formation and much of the Kirkstile Formation are locally intensely folded by synsedimentary slump folds. These folds range in size from a few centimetres to ten metres or more and are typically recumbent to isoclinal. Typical slump folds, which occur west of Gasgale Gill (NY 16412095), are disharmonic and commonly bounded by bedding parallel shear planes. The slump and gravity-slide folds are mainly overturned to the south-east. These folds have the main cleavage superimposed across them, some are tightened and others refolded by the later tectonic folds; examples occur on Gasgale Crags (NY 179221).

Tectonic folds and cleavages

On Whiteside End and Gasgale Crags, many examples of minor tectonic folds (F_1), with congruous axial-planar cleavages (S_1), occur. The axial planes and cleavages are mainly upright, northerly dipping (as at NY 17552211), but near the Gasgale Thrust they fan over to dip parallel to the fault, northwards at around 45°. Apart from the main cleavage (S_1), two other crenulation cleavages locally occur. One trends to the north or north-west and is mainly present as a lineation on bedding. The other is a late low-angle cleavage (S_2) with associated minor sideways-closing folds (F_2).

The Gasgale Thrust Fault

The Gasgale Thrust runs E–W along the foot of the main mass of Gasgale Crags. It dips northwards at 45–50° and thus has southerly-directed thrusting, and a throw of around 250 m. Here, the arenaceous lower part of the Kirkstile Formation is thrust over the less-competent siltstones higher in the forma-tion; the lithological change picks out the fault. The Gasgale Thrust has a similar direction of throw and is subparallel to the Causey Pike Thrust which post-dates the Crummock Water Aureole (Cooper *et al.*, 1988). It is probable that both faults have a long history of movement, both pre- and post-aureole, with the final displacement occurring after the Early Devonian.

The Crummock Water Aureole margin

At the west end of Gasgale Gill (NY 16502111) the gradational northern margin of the Crummock Water Aureole is present (Cooper *et al.*, 1988). This elongate E–W-trending aureole, dated at around 400 Ma (Cooper *et al.*, 1988), was produced by an unexposed, probably granitic body, possibly along a shear zone. At the west of Gasgale Gill (NY 16412095) the slump-folded siltstone of the Kirkstile Formation is bleached and hornfelsed; the colour changed from dark to light grey and the rock has a hard flinty appearance. Weathered surfaces here show the slump folding far better than unmetamorphosed outcrops.

Interpretation

Early interpretations of the structures in the present locality and the site at Hassness as being tectonic in origin are summarized in the description of the latter. In the present site it is clear that there are at least two generations of structures, one attributed to a soft-sediment origin during the Ordovician and one to protracted late-Caledonian (early Devonian) deformation.

The Gasgale–Whiteside area typifies the normal stratigraphical, sedimentological, and structural character of the Skiddaw Group north of the Crummock Water Aureole–Causey Pike Fault Line. In the two areas either side of this, the Skiddaw Group shows different sedimentological and early fold histories, but similar tectonic fold histories. The Gasgale–Whiteside area shows southerly derived Arenig Series greywackes and siltstones. These distal turbidite facies are folded by slump folds (F_0), overturned in a south-easterly direction (Webb and Cooper, 1988, their F_1). This over-turning is contrary to the source direction of sedimentation, and contrary to the westerly, over-turned, gravity-slide structures developed further south (see Hassness). These early structures, along with later, open folding, pre-date the Borrowdale Volcanic Group (Llandeilo–Caradoc) and probably relate to strike-slip movements along the Crummock Water–Causey Pike Line. This line, which separates the opposed overturning directions of the slump folds of Hassness and Gasgale, may approximately mark the axis of a local depositional basin.

Caledonian structures are represented by the main ENE- to NE-trending folds and associated cleavage (F_1 and S_1), the thrusting along the Gasgale Fault and the late sideways-closing folds with low-angled cleavage (F_2 and S_2). The Crummock Water Aureole is also Caledonian; dated at *c.* 400 Ma, it post-dates the D_1 structures, but pre-dates the D_2 structures (Cooper *et al.*, 1988). It is post-dated and bounded by a southerly directed thrust at Causey Pike, with which the Gasgale Thrust might be synchronous. The latest movements on these thrusts therefore post-date D_1 but they may well have had a long history of movement, the early parts of which may have been related to the evolution of the sedimentary basin.

Conclusions

The Gasgale–Whiteside area demonstrates that some structures which have previously been interpreted as tectonic in origin were produced by slumping and the gravity-driven sliding sediments. The contrast of the westerly movements of the gravity folds at the Hassness site with the south-easterly movements at the present site, suggest that they lay on opposite sides of the local depositional basin in earliest Ordovician times. The site shows the local refolding of these slump folds by tectonic folds, with their associated cleavages (closely spaced fine parallel fractures), both refolding and cleavage being the product of the Early Devonian Caledonian Orogeny. The area also illustrates the relationships of these structures to the metamorphism at the margin of the Crummock Water Aureole and thrust movements on the Gasgale Fault. The aureole, the baked and chemically altered zone of rock caused by the emplacement of the igneous intrusion, was formed around 400 million years ago towards the end of the Caledonian mountain building episode. The Gasgale Fault is even later, having moved at the very end of the orogeny, although it may have its origins in the evolution of the sedimentary basin.

RIVER CALDEW (NY 331325–325328)
D. E. Roberts

Highlights

The River Caldew section exhibits one of the finest sets of fold structures in the country. They are displayed with a clarity rare in the Ordovician Skiddaw Group, partly as a result of hornfelsing; they provide a critical locality for the understanding of the early deformation history of the group.

Introduction

At least two phases of deformation are represented here; the first producing originally sideways-closing N–S folds, but now characterized by a steep plunge, thought to be of slump origin, and the second the main end-Caledonian structures.

The section extends for almost 500 m upstream of a small dam (NY 331325) on the River Caldew and also includes the lowest seventy metres of Grainsgill Beck up to the contact of the hornfelsed

Skiddaw Slate Group with the greisen at the margin of the Skiddaw Granite. Although now hornfelsed, by the intrusion of the Skiddaw Granite, the rocks of the area have been identified as the Mosser–Kirkstile Slates (now Kirkstile Formation) (Arenig Series) (Jackson, 1961), equivalent to the Slates and Sandstone Group (Eastwood *et al.*, 1968). Roberts (1973, 1977a) favoured Jackson's usage, with a subdivision into a lower slate and sandstone division and an upper slates division. It is the lower division that is present in the River Caldew section, essentially a grey striped siltstone sequence with some thin mudstone layers and a few thin sandstone bands. The site displays a complex set of folds which have been significant in determining the structural evolution of the Skiddaw Group. The structures in the Skiddaw Group of the Northern Fells indicate four phases of deformation (Roberts, 1977a).

Some aspects of the site are referred to in the Memoir of the Geological Survey (Eastwood *et al.*, 1968); various opinions of the structures are mentioned in a report of a Geologists' Association field meeting (Mitchell *et al.*, 1972); but the definitive paper on the section is that of Roberts (1971), which formed part of a fuller study of the Skiddaw Group of the Northern Fells (Roberts, 1973, 1977a).

Description

A map of the section is presented herein (Figure 3.9) and it shows the overall structure, together with offset diagrams of key localities with accompanying stereographic projections.

The style of folds in the section is complex, with frequently more than one style being present in a single fold. Folds show concentric, similar, chevron, and conjugate fold elements, although close to tight similar folds are the most dominant style (Figure 3.10). Interference patterns caused by refolding can also be detected, a good hook-type being visible at the junction of Grainsgill Beck (see Figure 3.9; Locality A). One common feature of many structures is disharmony of fold wavelength in the core of the folds, together with discordant structures. There is no obvious consistent orientation of the structures, although a generalized NE–SW trend can be detected. What is obvious, however, is the steep plunge of the folds over the entire section, with only a few exceptions. Roberts (1971) removed a regional dip from these

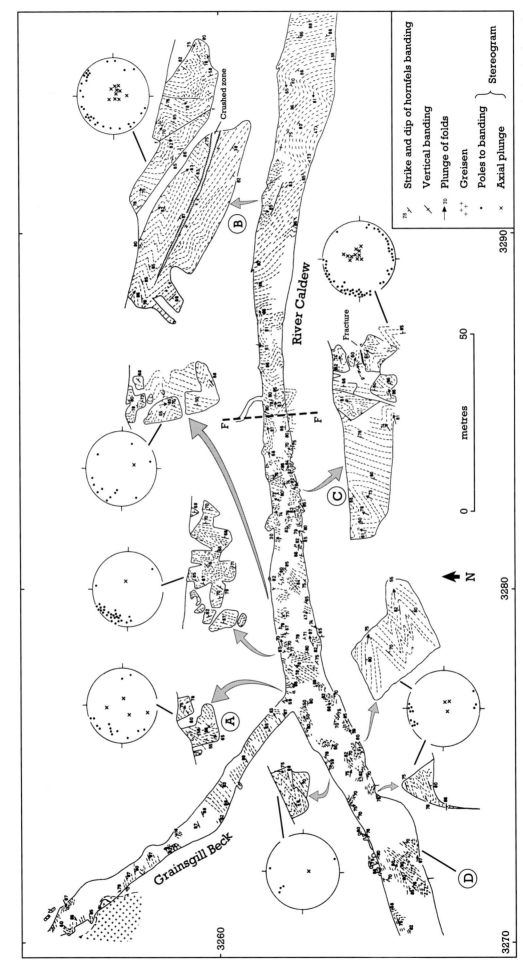

Figure 3.9 Structural map of the River Caldew section, showing the main folds in the hornfelsed slates of the Skiddaw Group. Inset diagrams are enlargements (×4) of selected localities, each with an accompanying equal-area projection of the structural elements of that locality (after Roberts, 1971). Localities A–D referred to in text.

Figure 3.10 River Caldew. Steep NE-plunging folds in hornfelsed slates of the Skiddaw Group produced largely by gravity-driven slumping. They show the truncations and variable, disharmonic style typical of this process. Bedding planes are chalked (compass top right for scale). (Photo: D. Roberts.)

structures and showed that originally many had a N–S recumbent attitude. Good representatives of the overall nature of the structure are seen at several localities: a prominent set of folds near the road approximately 200 m west of the dam (Figure 3.9, Locality B); exposures each side of a small waterfall 30 m west of the small tributary (Figure 3.9, Locality C); and those 50 m upstream of the confluence of Grainsgill Beck (Figure 3.9, Locality D).

One important aspect of these folds, however, is that despite their tight nature, often angular hinges, and the presence of such structures as conjugate fold pairs, many folds show evidence of disruption, convolute folds, and some disharmony not consistent with tectonic structures. It is the interrelationship of structures which appear to be tectonic, with others that are more consistent with soft-sediment deformation, that makes this particular site of special interest. The absence of any tectonic fabric, such as cleavage, due to the effects of metamorphism, adds to the problems of establishing the true nature of the deformation history.

Interpretation

The majority of structures in the River Caldew Section can be precluded from late-Caledonian deformation events, on account of their steep plunge and variable orientation, which is totally inconsistent with any known main-phase structures in the Lake District. Consequently, it is generally accepted that many of the folds in this section are the product of an earlier deformation phase of the Skiddaw Group.

The suggestion of Brown *et al.* (1964), that the tight folds in the hornfels were formed as a result of the forcible injection of the granite, is also rejected, on the grounds that these folds are earlier than the main late-Caledonian (D_1) deformation, whereas the granite was probably intruded either synchronous with, or slightly later than D_1, but certainly earlier than D_2 (Soper and Roberts, 1971).

Roberts (1971) preferred to attribute these structures to pre-Windermere Group deformation, on account of their original N–S trend which was consistent with that reported for the 'pre-Bala' (pre-Caradoc) folds in the Borrowdale volcanics (Mitchell, 1929). However, as outlined in the introduction to this chapter, detailed study has shown that north-trending pre-Caradoc folds do not exist within the Borrowdale Volcanic Group (Soper and Numan, 1974). Moreover, Branney and Soper, 1988 have demonstrated that deformation in the Borrowdale Volcanic Group and at the Borrowdale–Windermere Group unconformity originated from volcanotectonic, rather than compressive, processes. A resurvey of the west section of the Skiddaw Group has demonstrated that the north-trending folds are most probably gravity or slump structures (Webb and Cooper, 1988) (F_0 of this volume).

A close re-examination of the folds in this section reveals much disharmonic folding, disruption in the fold hinge, apparent shears sealed prior to metamorphism, and some irregular convolute folds not dissimilar to features produced as a result of slump folding of unconsolidated sediments. This, together with the original recumbent attitude, would point to gravity sliding of unconsolidated material as a more likely mechanism for the formation of the folds in the Caldew Valley, as suggested by Roberts (1977a). It is not at all easy to put an indisputable age to the formation of the folds, but regional evidence would point to instability during the onset of Llanvirn or Llandeilo Series volcanism as being the most likely cause. Since Llanvirn volcanic rocks occur to the north of Carrock Fell, this is the favoured time for fold formation; the proximity of those rocks may be the reason why folds are so prolific in the Caldew Valley but are much less common elsewhere in the Lake District.

Perhaps it would be unwise to categorically state that all the folds in the River Caldew section are the result of gravity sliding, since it is highly likely that these structures were subsequently deformed by main-phase Caledonian deformation. This could account for the refolds and also for those structures with a gentle plunge to the east or west. The tight nature of some folds with angular hinges is also likely to be the result of the recumbent folds having been flattened by the main Early Devonian tectonic event. However, in view of the complexity of gravity folding, which itself could produce refold patterns, it would be a difficult task to distinguish individual structures which could conclusively be attributed to main-phase Caledonian deformation.

Conclusions

The exposures of baked (hornfelsed) Skiddaw slates in the River Caldew section display a complex set of folds which can best be interpreted as the original product of the sliding of unconsolidated sediments under gravity. This first deformation event (D_0) is likely to have occurred during the onset of volcanism in Llanvirn times, that is, during the early part of the Ordovician Period. The original flat-lying N–S folds were subsequently refolded to some extent during the main phase of Caledonian tectonic movements (D_1), in the Devonian. During these movements, folds were tilted steeply to the north, thus accounting for the steep attitude (plunge) that the fold hinges now display. Earlier suggestions that the original N–S folds were the result of Late Ordovician (Caradoc) tectonic deformation are now completely rejected. The site has been of interest for a long time, and, although the interpretations of the structures have changed, the remarkable clarity with which the folds are displayed makes it one of the most significant sites in the Lake District for understanding the development of the Caledonides in that area.

RAVEN CRAGS, MUNGRISDALE (NY 363306–360311)
D. E. Roberts

Highlights

Raven Crags provides one of the best-displayed examples of Caledonian folds in the Lake District Skiddaw Group. At least four episodes of crustal deformation can be recognized here. These outcrops expose some of the most complex fold structures documented in the region, and they have produced important information on the sequence of events during the Caledonian Orogeny.

Introduction

The site extends for some 500 m north of School-House Quarry, Mungrisdale, on the eastern flank of the Northern Fells. The crags themselves comprise a set of vertical and steeply inclined exposures, separated by relatively flat ground with no exposures. They expose rocks which have been assigned to the Loweswater Flags division (now Loweswater Formation) of the Skiddaw Group

(Jackson, 1961) and are, here, a part of a small triangular-shaped fault-bounded inlier almost 2 km long, the eastern margin of which is the major Carrock-End Fault. A number of prominent hollows more or less perpendicular to the crags themselves represent the lines of faults or master joints.

The Loweswater Formation is a group of alternating greywacke sandstones, siltstones, and slates. On many of the crag exposures, small-scale sedimentary structures are clearly visible on weathered surfaces in the more arenaceous beds, indicating their origins as turbidity flows.

The sections reveal three sets of structures, which post-date the D_0 folds described above in the description of the Caldew River site. In this account they are labelled D_1, D_2, and D_3 (but see Table 3.1 for other attributions). The D_1 structures dominate the crag sections; upright folds with steeply inclined slaty cleavage generally have anomalous north-westerly trends. The D_2 structures comprise subhorizontal folds and gently inclined cleavage, whereas D_3 structures, only recorded at the south-west end of the crags, are minor N–S flexures with rare fracture cleavage.

Some aspects of the site are referred to in the Memoir of the Geological Survey (Eastwood *et al.*, 1968). Various opinions of the structures, now superseded, are mentioned by Mitchell *et al.* (1972), but the definitive paper on the section is that of Roberts (1977b) which formed part of a fuller study of the Skiddaw Group of the Northern Fells (Roberts, 1973, 1977a).

Description

The general structure of Raven Crags is shown in Figure 3.11. They can be divided into a southern section where the structure follows the regional E–W trend, for approximately 100 m to the north of the quarry, and a northern section, where asymmetrical folds with a gentle WSW limbs and a steep ENE limbs have an anomalous north-westerly trend. The plunge of the folds, which can be traced for distances up to 100 m, is gently inclined to the north-west and the axial planes are steeply inclined to the south-west.

At School-House Quarry, Mungrisdale (NY 363306), black slates containing *D. deflexus*, (Jackson, 1961) dip steeply to the south, and have a weak S_1 E–W cleavage, subparallel to the bedding. Two gently inclined E–W thrusts, with associated subhorizontal folds, dominate the struc-

ture, and a faint crenulation parallel to the thrust is also visible on bedding planes over most of the quarry. These are all D_2 structures and the thrusts are most likely to be accommodation structures for stresses acting during the formation of the subhorizontal open D_2 folds. At the eastern end of the quarry, two highly altered quartz-dolerite dykes occur parallel to the bedding and are visible, both in the side wall and in the quarry floor where they have been displaced along minor N–S faults. Steeply plunging N–S minor folds, with a steeply inclined associated fracture cleavage, occur in the slates adjacent to the dykes along the lines of the faults, and these are regarded as being D_3 structures. The dykes reacted to the D_3 stresses by brittle fracture, whereas the more ductile slates accommodated the stresses by folding and the development of the fracture cleavage.

Immediately to the north-west of Mungrisdale Quarry, D_1 structures dominate, with, in parts, a strong overprint of D_2. The D_0 structures are not common and, although they follow the D_1 trend, can be distinguished from them by their steep plunge. D_3 structures are also uncommon outside School-House Quarry. Good examples of asymmetrical folds, with a NW–SE trend, occur over the entire crags. One hundred metres to the north-west of the north-east end of Undercrag Farm, at the base of the crags, a prominent, upright open syncline occurs with associated minor folds and a cleavage. This last approximates to a true slaty cleavage in the mudrocks, but with an accompanying cleavage refraction it becomes a spaced cleavage in the sandstones.

The anomalously trending folds are identified as D_1 structures on account of their similarity in style to D_1 structures elsewhere in the Northern Fells, their associated, well-displayed, slaty cleavage, and the gradual change in their trend from W–E to NW–SE immediately to the west.

Interpretation

At this locality, structures occur that are clearly attributable to four deformation phases, related to the Caledonian Orogeny. However, the anomalous trend of structures assigned to the main structural event is something of a problem, since it is NW–SE compared with NE–SW for the Caledonides as a whole, and W–E for the same structures in the surrounding area.

The author at one time, and others (Mitchell *et al.*, 1972), considered that these folds were the

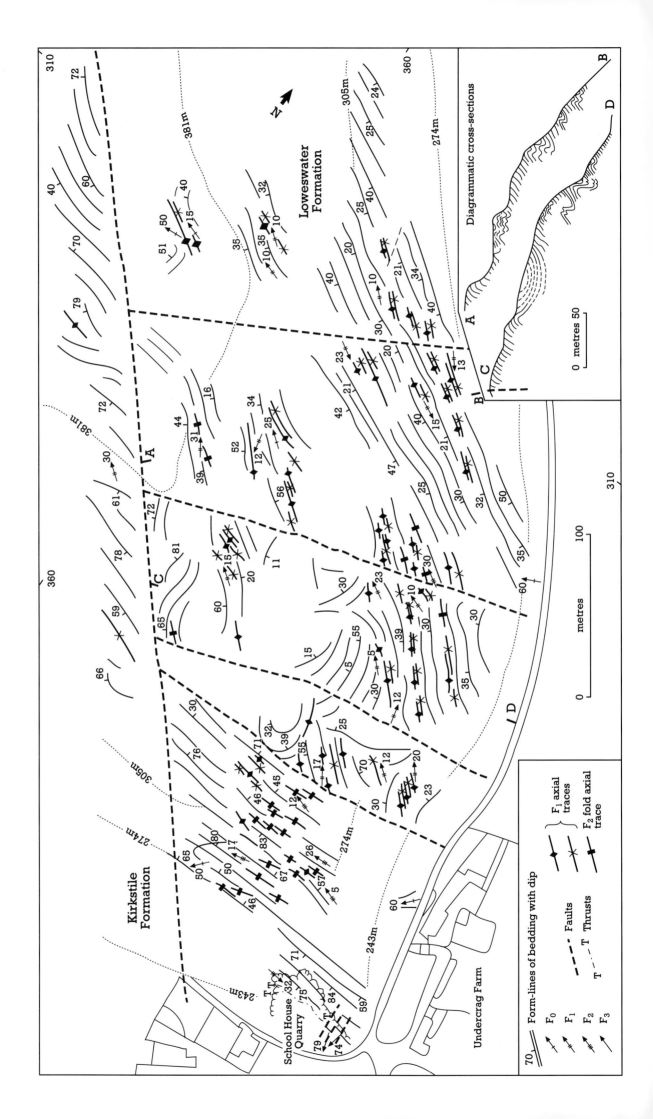

result of an early-Caledonian deformation phase of the Skiddaw Group. Early deformation structures (D_0) are clearly visible in the nearby Caldew Valley, but the folds on Raven Crags bear no resemblance to these in either style or orientation. These folds have a gentle plunge, whereas D_0 folds have a steep plunge (as is seen locally at this site), and their associated cleavage is typical of S_1.

It was considered that the anomalous NW–SE-trending folds at Raven Crags might be the Skiddaw Group equivalent to supposed north-trending folds affecting the Borrowdale Volcanic Group. However, as outlined in the introduction to this chapter, such folds do not exist (Soper and Numan, 1974) and Borrowdale Group deformation is related to volcanotectonic processes (Branney and Soper, 1988). Roberts (1977b) believed these folds to be late-Caledonian D_1 structures, and he suggested three alternative causes for their anomalous trends:

1. A continued arcuate swing from NE–SW to E–W to NW–SE across the Lake District; against this suggestion are E–W structures further east at Troutbeck (NY 385270) and even at the southern end of the crags themselves in School-House Quarry.
2. Drag associated with the right lateral movement along the Carrock-End Fault. However, the NW–SE trend is not consistent along the length of the fault and so this suggestion should also be rejected.
3. Reorientation during the D_3 phase. Minor refolding of earlier structures, by the D_3 event, is a common feature throughout much of the Northern Fells, but it is generally on a minor scale. The suggestion is that the more competent Loweswater Formation within the fault bounded inlier was re-oriented as a mass, whereas minor D_3 structures were formed in the less-competent slates. Roberts (1977b) considers this to be the most likely explanation, although it does not explain why such a large-scale change in trend has not been recorded in competent Skiddaw Group rocks elsewhere in the area. To the south-east, across the drift-covered plain, the volcanic rocks of Eycott Hill have a N–S strike, and there may be some connection between this and the trend on Raven Crags, but that has still to be established.

Conclusions

The exposures of the Loweswater Formation, in the fault-bounded inlier of Raven Crags, show, with remarkable clarity, some of the most complex fold structures documented in the region, and they have provided important contributions to an understanding of the Caledonian Orogeny in north-west England. Four phases of deformation are recognized (D_0 and D_{1-3}): the first (D_0) the product of slumping and crumpling of sediments on a sloping Ordovician (Arenig) sea-bed; the second, the dominant regional structure (probable Devonian–D_1); the third (D_2), dominant here, includes the thrust faults in the site, and the fourth formed next to a pair of dolerite dykes which are also displaced by some later, small faults.

Here there was clearly a long history of strain and deformation in the form of crustal shortening and folding, and also faulting. These are some of the most complex and informative outcrops in the region, providing graphic evidence of the length and intensity of the Caledonian mountain-building episode. The main Caledonian phase has a NW–SE trend which is anomalous to that of other tracts of Skiddaw Group rocks in the Lake District.

WARNSCALE BOTTOM, BUTTERMERE (NY 201135; NY 199135)
F. Moseley

Highlights

The site provides a rare opportunity to examine evidence for the nature of the intra-Ordovician unconformity between the Skiddaw Group and the Borrowdale Volcanic Group. Tight folding and cleavage in the former, contrast dramatically with the uniformly dipping Borrowdale lavas above. The faulted and depositional junctions between the two groups gives evidence of their true stratigraphical and structural relationship, which has been the subject of much contention in the past.

Figure 3.11 Map of the structures in the Loweswater Formation on Raven Crags, Mungrisdale. A–B and C–D are the lines of the cross-sections illustrated in the inset (modified from Roberts, 1977b).

Introduction

Ever since geologists began to make detailed maps of the Lake District, there has been controversy about the junction between the slates of the Skiddaw Group and the Borrowdale Volcanic Group, whether it was conformable, unconformable, or faulted (see Moseley, 1972 for review).

Some exposures of the junction (such as in the present site) are clearly faults, but interest has focused, in recent years, on whether the junction was originally a major orogenic unconformity. Simpson (1967) proposed that two phases of deformation preceded the volcanics, the latter being affected by only gentle folding and a single cleavage. Soper (1970) challenged this interpretation showing that, where the junction is exposed, a single cleavage in the slates passes into the overlying tuffs, and that slate fragments in the tuff show a common cleavage. In spite of various arguments for and against the hypotheses (see 'Introduction', Chapter 1) Soper's (1970) view has prevailed, albeit with much modification.

Several workers have noted that significant north-trending folds, with no associated cleavage in the Skiddaw Group, do pre-date the volcanics (for example, Roberts, 1971, 1977a; Jeans, 1972; Wadge, 1972; Webb, 1972). It now seems to be agreed that these folds (F_0 elsewhere in this volume), which may be tight and have amplitudes up to 500 m, are the product of submarine slumping. These folds are of variable trend. However, since the top of the Skiddaw Group ranges in age, from Upper Llanvirn in the east to Arenig in the west, where it is overlain by the Borrowdale Volcanic Group (Soper and Moseley, 1978), a regional unconformity certainly exists. Several authors and, more recently, Webb and Cooper (1988) have related the unconformity to the incipient Lake District Anticline (Downie and Soper, 1972), and Branney and Soper (1988) have associated the unconformity with both slumping and volcanotectonic uplift.

Description

Two localities are described, the first in Warnscale Beck, the second in Black Beck to the west. The first (NY 201135) (Moseley, 1975; Wadge, 1978a) provides a continuous section across the junction of the Skiddaw and Borrowdale Groups (Figure 3.12). Starting downstream, the Skiddaw Group is banded with pale, silty layers in dark pelite, and the cleavage is moderately strong, being visibly axial planar to small folds, but also often subparallel or parallel to the silty laminae. Along the stream bottom, the bedding in the slates is clearly seen but not the cleavage. On the stream banks, weathering has clearly picked out the cleavage, but the silty laminae can be seen to be tightly folded as far as the junction, where these folds are abruptly truncated by a sharp plane inclined 60°SE, which must be regarded as a fault. Continuing upstream, a massive, flow-jointed but unfolded andesite dips steadily to the south-east.

Black Beck (NY 199135) shows a 20 m exposure across the Skiddaw–Borrowdale Groups junction, and although the contact is not so clear, the bedding in the slate is discordant to the junction, which does not appear to be faulted (Bull, unpublished; Wadge, 1978a). Wadge reported 2.8 m of conglomerate at the base of the volcanics, and Bull noticed that there were slate blocks near the base, and that the cleavage in them was parallel to that in the underlying Skiddaw Group (60/120°). Bedded tuffs, some distance above this locality, show the development of a strong cleavage with essentially the same attitude as that in the Skiddaw Group below.

Interpretation

The arguments that have centred on the nature of the Skiddaw Group–Borrowdale Volcanic Group junction can be easily appreciated at these two localities. Below the junction are strongly cleaved, tightly folded sediments; above are uniformly dipping massive volcanics, with only locally a crude cleavage. These observations, in themselves, might only indicate the contrasting behaviour of incompetent and competent lithologies, although the persistence of minor folds in the Skiddaw Group close to the junction and the presence of a conglomerate and of mudstone clasts in the volcanic sequence do suggest an unconformity. The presence of a single cleavage common to both slate and tuff clearly shows that any unconformity pre-dates substantial shortening in the rocks.

The significance of the folding in the Skiddaw Group is problematical. Unlike some folds observed near this junction, they have a Caledonoid north-easterly trend, with axial-planar cleavage. In the light of the recent work of Webb and Cooper (1988), however, it is considered that the folds are slump structures with an original north-easterly trend, which have been tightened, together with

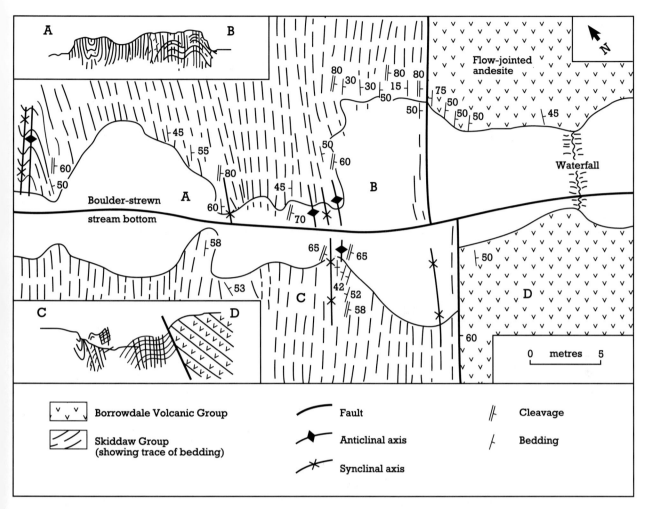

Figure 3.12 Detailed map and sections of the Skiddaw Group–Borrowdale Volcanic Group junction exposed in Warnscale Bottom. Anticlines, synclines, dip of bedding and cleavage are shown (after Moseley, 1975).

the formation of cleavage, during the late-Caledonian deformation. This origin would explain the persistence of the folds close to the unconformity and at the same time emphasize the importance of slumping in its evolution (Webb and Cooper, 1988).

Conclusions

These exposures provide evidence for the nature of the unconformity between two of the major stratigraphical units in the Lake District; that relationship has been the subject of intense debate in the past. There have been arguments over whether the junction was an unconformity and whether this unconformity was evidence of major earth movements, and how far it was affected by faulting. It was once thought that folds in the

Skiddaw Group rocks were evidence of early-Caledonian deformation that pre-dated the eruption of the Borrowdale volcanics. However, by comparison with folds at other sites (discussed above), it is now clear that such folds were formed soon after deposition of these Skiddaw sediments on the sea-bed. The relationships seen suggest that the unconformity, while real, is not of orogenic proportions as was once proposed, and it has been locally modified by faulting.

HOLLOWS FARM (NY 245170)
F. Moseley

Highlights

This site provides unique continuous exposures across the critical junction between the Skiddaw

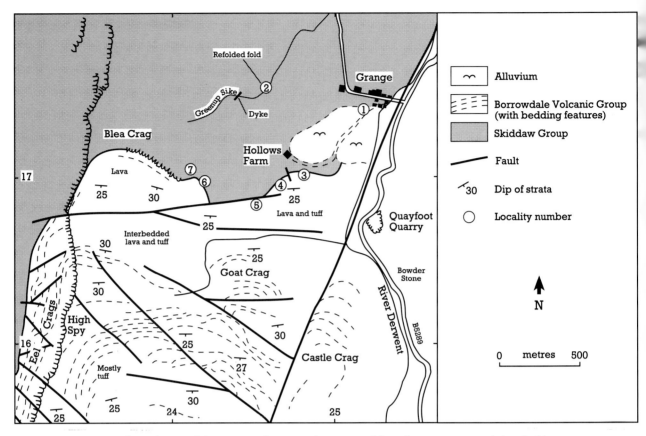

Figure 3.13 Geological map of the junction between the Borrowdale Volcanic Group and the Skiddaw Group in the area around Grange-in-Borrowdale, Cumbria, showing localities mentioned in the text.

Group and the Borrowdale Volcanic Group. Many of the erosional features associated with the unconformity are seen, but the main Caledonian cleavage post-dates these features and affects both Groups.

Introduction

The exposures around Grange and Hollows Farm are important in the discussion concerning the Skiddaw Group–Borrowdale Volcanic Group junction (Soper, 1970; Jeans, 1971, 1972; Mitchell *et al.*, 1972; Wadge, 1972). Soper (1970), especially, used these exposures to refute the hypothesis of Simpson (1967), that this junction was a large-scale unconformity, the product of orogeny. A defence of the latter position, based upon these exposures has been presented by Mitchell *et al.* (1972, pp. 455–8). The general arguments concerning this junction are set out in the introduction to this chapter and in the decription of the Warnscale Bottom site.

Description

At several localities (1–7 of Figure 3.13) the relationships at the Borrowdale–Skiddaw junction are seen in complementary illustrative outcrops.

Locality 1: At Grange (NY 253175) Skiddaw slates are exposed on a glacially smoothed slab adjacent to the river. The bedding is tightly folded with a N–S trend, but with no obvious cleavage. It is likely that these folds are slumps.

Locality 2: In Greenup Sike (NY 246176) there is a thin sandstone band in slate which reveals a fold with vertical plunge. This is believed to be a slump fold which has been deformed by the late-Caledonian deformation (D_1).

Locality 3: The classic locality in Scarbrow Wood (NY 249170) exposes the Skiddaw Group–Borrowdale Volcanic Group junction, and has been controversial for over 100 years (see Mitchell *et al.*, 1972). Soper (1970, Locality B) reported that here andesite tuff rests on an eroded surface of a conglomerate of mudstone fragments. The tuff also contains mudstone fragments and a single cleavage affects all rocks, including the underlying Skiddaw slates, which itself is tuffaceous. It is now suggested that a minor unconformity can be seen here, but also that the minor folds in the Skiddaw slates, in adjacent sections, are due to slumping and not to tectonic activity – see the Introduction to this chapter. Nearby, Soper (1970, Locality C) reported a tuff-filled channel in the underlying conglomeratic mudstone, and that cleavage passes across the junction.

Locality 4: On the fellside immediately above Scarbrow Wood (NY 247170), observations may be made which are similar to those made at Locality 3 – see above. Soper (1970, Locality D) observed here, that the S_1 cleavage is deformed by F_2 folds, which have horizontal axial planes and a gentle north-easterly plunge.

Localities 5, 6, and 7: Adjacent to these outcrops (Locality 5, NY 245169), gently dipping andesitic tuffs are faulted against the Skiddaw Group. Higher up the fellside below Blea Crag (Localities 6 and 7, NY 242170 and NY 241171) there are other exposures of the junction, which here consists of a thin conglomerate made up of Skiddaw mudstone pebbles resting on Skiddaw mudstone (Soper, 1970, Locality F). Bedding in the pebbles is disturbed, but cleavage has a common attitude in pebbles and matrix.

Interpretation

The interpretation of this site relies on criteria which are very similar to those used at Warnscale Bottom – see above. However, the principal attraction of this site is that it appears to provide, albeit poorly exposed, continuous sections across the Skiddaw–Borrowdale junction. It is reported (see Mitchell *et al.*, 1972, p. 457) that the slate itself is tuffaceous and that continuous bedding is difficult to define within a few metres of the first true volcanics. The topmost mudrocks appear, in places, to comprise a conglomerate of mudstone fragments, with no consistent orientation, set in a mudstone matrix. Locally, the lowest tuff contains mudstone fragments. A difficulty seems to be the identification and placing of any single surface of unconformity, although local erosional surfaces at the base of tuff horizons are claimed.

The site also appears to provide evidence that the principal cleavage in the pelites passes continuously into or, at least has the same attitude as, the tuffs above. This must be strong evidence that

whatever the nature of the unconformity, it did not post-date major cleavage-related deformation. The cleavage common to both the Skiddaw and Borrowdale Groups is that of the main end-Caledonian deformation. Against this evidence, as at Warnscale Bottom, is the presence of tight, locally overturned, minor folds in the bedded mudrocks, which are absent in the tuffs above. Soper (*in* Mitchell *et al.*, 1972, p. 456) argued that this was due to the inability of the competent tuffs to develop minor folding comparable with that in the Skiddaw Group, rather than pre-volcanic folding. However, Soper (*in* Branney and Soper, 1988) now accepts the recent proposal by Webb and Cooper (1988) that such folds resulted from slumping.

Conclusions

These exposures provide important evidence for the nature of the unconformity between two of the major stratigraphical units in the Lake District. Unique continuous sections across the junctions show it to be transitional, although marked by various erosional features, which may be related to the initiation of volcanicity and slumping. There appears to be no evidence that the unconformity was related to an early compressional phase during a mountain-building episode, as was once suggested.

It is now assumed, on the basis of the fact that both the Skiddaw Group and the Borrowdale Group volcanics share the same cleavage pattern (fine, closely spaced, parallel fractures), that Caledonian deformation events are much younger than the age of the unconformity. Therefore the unconformity, once assumed to be evidence of early Caledonian earth movements during the Ordovician, is now taken to represent lesser-order intra-Ordovician movements and folding in the Skiddaw Group. Caledonian folding and cleavage was superimposed much later probably during the Devonian.

LIMESTONE HAWS TO HIGH PIKE HAW, CONISTON (SD 279966–255940)
F. Moseley

Highlights

This site provides the only location for demonstrating the important folding that occurred at the end of mid-Ordovician volcanicity. Folded Borrowdale Group volcanics are spectacularly overstepped by the later sediments. The locality also exposes late-Caledonian strike-slip faults, which abruptly turn into thrusts along incompetent shales.

Introduction

This classic area includes outcrops in the upper part of the Borrowdale Volcanic Group and the overlying sedimentary rocks, now assigned to the Windermere Group (Ordovician–Silurian). The extinction of the Borrowdale 'volcano', an event almost certainly related to partial closure of the Iapetus Ocean, near the end of the Ordovician (Williams, 1975; Moseley, 1977), was followed by, or was contemporaneous with, the intrusion of the Ordovician component of the Lake District batholith (Firman and Lee, 1986; Soper, 1987). It is likely that these events resulted in uplift, folding, and erosion before the Late Ordovician Coniston Limestone was laid down unconformably upon the volcanics.

The best documented of these pre-Coniston Limestone Formation folds is the Ulpha Syncline, recognized by Aveline *et al.* (1888), Mitchell (1956a) and Firman (1957), mapped by Numan (1974) and discussed by Soper and Numan (1974), Soper and Moseley (1978), and, more recently, by Branney and Soper (1988). This major E–W trending fold has assumed some importance as being a indication of Late-Ordovician compression, clearly pre-dating the late-Caledonian deformation represented by cross-cutting cleavage. Branney and Soper (1988), however, have removed the effects of later cleavage-related deformation and rotated the fold limbs to their pre-Coniston Limestone attitudes. The resulting fold is a weak monoclinal flexure which they consider more compatible with bending that would have been associated with foundering of the volcanic pile.

The final closure of Iapetus resulted in continental collision, and strong folding and cleavage across the whole of the Lake District. The major fold, affecting the Coniston area, was the Wrynose anticline, the south-east limb of which extends 8 km from Wrynose Pass to Coniston Water and has resulted in the steep south-easterly dips in both the Borrowdale Group volcanics and the lower part of the Windermere Group (Mitchell, 1940; Soper and Moseley, 1978) seen here.

Subsequent to the folding, the latest adjustments to the closing of Iapetus were by strike-slip faulting, often sinistral in displacement (Soper and Hutton, 1984). In the Lake District (Moseley, 1972), the large faults, such as the Coniston Fault, are north-trending with sinistral displacement, but an important north-west-trending set with dextral displacement are particularly well seen along the Borrowdale–Windermere Group boundary. Many of these faults turn into low-angle thrusts above the volcanics, particularly utilizing the fissile black shales of the Skelgill Beds.

Description

Faulting near Limestone Haws and the Walna Scar Track (SD 279965 and Figure 3.14A)

The volcanic structures, in this area, are difficult to determine, the lithology being mostly ignimbritic breccia of probable laharic origin (Yewdale Breccia Formation), with little indication of the dip. The breccia is overlain by the Long Sleddale Member, a tuffaceous sandstone with occasional brachiopods, followed by members of the Coniston Limestone Formation (Moseley, 1983, 1984). At High Pike Haw, a fold (the Ulpha Syncline) in the Yewdale Breccia is overstepped by the Coniston Limestone. Dips in the latter are steep, generally between 60° and 80° to the ESE. This sequence is overlain, discordantly, by the Skelgill Member (Llandovery Series), and could easily be interpreted as an unconformity. However, the Skelgill Member, where well exposed, is seen to be much thicker than the space available in the narrow marshy gully between the Coniston Limestone Formation and the Browgill Beds (Figure 3.14A), and this implies the presence of a strike fault. The Skelgill Member comprises black, highly incompetent graptolitic mudstones, known from many outcrops across the Lake District to be followed by thrusts, subparallel to bedding. The discordance, therefore, is attributed to thrusting rather than to an unconformity. It is noticeable that the dip faults, clearly seen above the Walna Scar track, probably small wrench faults, terminate at this horizon and the suggestion is that they are linked wrench-thrust faults (Soper and Moseley, 1978).

Faulting from Flask Brow (SD 270960) to Ashgill Quarry (SD 269954) (Figures 3.14B and 3.14C)

In the area between Torver Beck and Ashgill Quarry, there are approximately ten dip faults which displace the Borrowdale Group–Coniston Limestone Formation junction. They have north-westerly trends and are likely to be dextral wrench faults with strike-slip displacements of up to 100 m. Most of the faults do not cross the Skelgill Member outcrop but are believed to rotate into thrusts, as described above. Between the Torver Beck tributaries (SD 276962) and Ashgill Quarry (Figures 3.14B and 3.14C; Moseley, 1983) the junction between the Borrowdale Volcanic Group and the Applethwaite Member (Coniston Limestone Formation), can be seen to be displaced. The fault at the south-west end of Ashgill Quarry is, however, different in that it does, also, displace the Skelgill Member, by some 70 m (Figure 3.14C). A strike fault seen in the quarry (Figure 3.14B) also displaces the member.

The Caradoc unconformity at High Pike Haw

At SD 260950, the Coniston Limestone Formation (Windermere Group) oversteps the Ulpha Syncline – see Figure 3.15; based on Soper and Numan (1974) and Branney and Soper (1988). In so doing, it cuts across more than 600 m of the Borrowdale Volcanic Group, including the Yewdale Breccia, Yewdale Bedded Tuff, and the Tilberthwaite Tuff Formations (Mitchell, 1940; Soper and Moseley, 1978; Soper, 1987; Branney and Soper, 1988). Figure 3.4 shows form lines for bedding in the principal areas of exposure in the volcanics, and strike and dip values for exposures in the clearly unconformable Coniston Limestone Formation.

The north limb of this fold strikes 045°, with moderate to steep dip south-east, while the south limb has an average N–S strike, dipping moderately east, with variations due to medium-scale folding. The fold has a plunge gently to the ENE, has an interlimb angle of 115°, an amplitude of 4 km and a half wavelength of 8 km. The cleavage also strikes ENE, some 20°–30° anticlockwise of the fold axial-plane trace. The overstepping Coniston Formation has a constant north-easterly strike and moderate south-easterly dip, clearly oblique to that in the Borrowdale Group volcanics (particularly around SD 265950), but it shares a common cleavage orientation with those rocks.

Interpretation

The Ulpha Syncline is the one fold structure in the Borrowdale Volcanic Group that can be clearly demonstrated to pre-date the Coniston Limestone Formation (Windermere Group). Cleavage of late-Caledonian age, equally clearly, cross-cuts both the fold and the unconformity. The phase of Late Ordovician folding demonstrated by the syncline has been credited with considerable importance in the history of the Lake District (Soper and Numan, 1974) and has been used to support the hypothesis that the closure of Iapetus essentially occurred in the Ordovician (Murphy and Hutton, 1986). The demonstration, however, by Branney and Soper (1988) that this fold was an open mono-cline, before Windermere Group sedimentation, has reduced its significance; it is now associated with Borrowdale Group volcanotectonic faulting and block-tilting connected with caldera collapse.

It is quite clear that most of the NW-trending faults that cut the Borrowdale Volcanic Group and the Coniston Limestone Formation do not displace the early Silurian Skelgill Beds and higher formations. From the relationships seen, it seems most likely that the faults must be linked strike-slip/thrust structures. Thus the site demonstrates that one of the pre-Coniston (Limestone Formation) fold structures in the Borrowdale Group is not a significant tectonic structure but rather of volcanotectonic origin. Some of the late, north-west-trending faults in this area terminate as thrust structures within the bedding.

Figures 3.14A, B, and C (on pages 80 and 81) Geological maps illustrating the nature of the faulting in three areas within the Limestone Haws–High Pike Haw, Coniston site (after Moseley, 1990, Figure 52B, C and D). (A) South side of Little Arrow Moor. (B) Area around Torver Quarry. (C) Area around Ashgill Quarry.

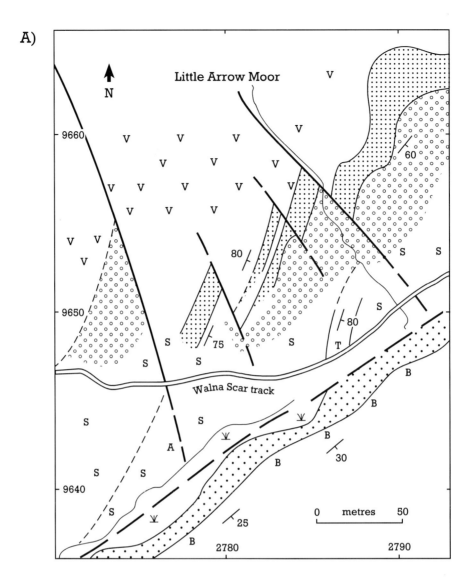

A)

Little Arrow Moor

N

9660

V V V V V
V V V V
V V V V
V V V V
V V V
V V
V

9650

S

S

S

80

75

S

S

S

T

80

60

S S

S

Walna Scar track

9640

S

S

S

S

S

S

A

B

B

B

30

B

25

B

0 metres 50

2780 2790

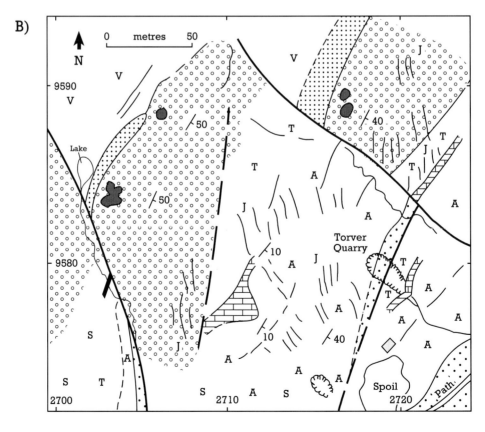

B)

N

0 metres 50

9590

V

V

V

Lake

50

50

V

T

T

A

J

40

J

T

T

J

A

A

9580

J

10

A

J

T

Torver
Quarry

T

A

A

A

S

A

J

10

40

A

A

A

A

Spoil

Path

S

T

S

A

S

A

A

A

2700 2710 2720

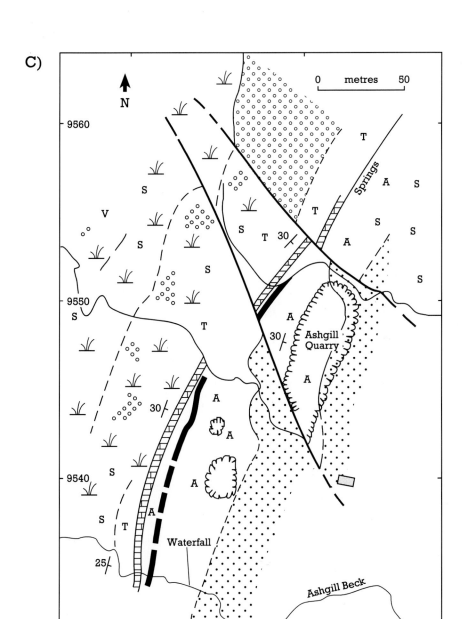

C)

9560

9550

9540

N

0   metres   50

Springs

Ashgill
Quarry

Waterfall

Ashgill Beck

2680   2690

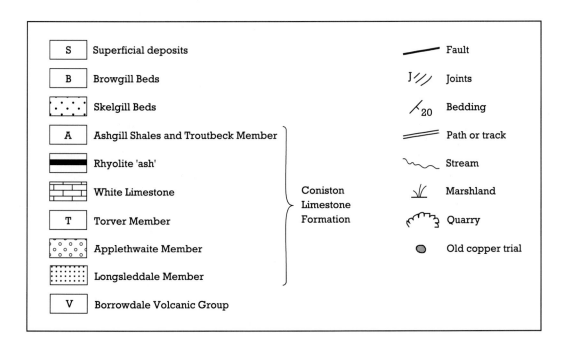

S	Superficial deposits	
B	Browgill Beds	
∴	Skelgill Beds	
A	Ashgill Shales and Troutbeck Member	⎫
▬	Rhyolite 'ash'	
▤	White Limestone	⎬ Coniston Limestone Formation
T	Torver Member	
° °	Applethwaite Member	
∵	Longsleddale Member	⎭
V	Borrowdale Volcanic Group	

▬▬	Fault
J ⫽	Joints
⤸20	Bedding
═	Path or track
∿	Stream
⅄	Marshland
⌓	Quarry
◉	Old copper trial

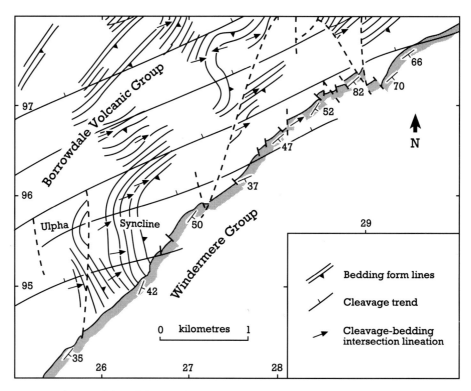

Figure 3.15 Structural map of the Ulpha Syncline at Torver High Common (after Soper and Numan, 1974; Soper and Moseley, 1978, figure 24).

Conclusions

The Ulpha Syncline is an important structure that was produced as the result of Ordovician volcanic events. It has considerable bearing on hypotheses regarding the history of closure of Iapetus Ocean. The principal locality is important as the only area where the unconformity of the Coniston Limestone Formation (Windermere Group) with the underlying folded Borrowdale Group volcanics can be demonstrated. Adjacent localities are important, not only for providing a rare opportunity to examine and demonstrate the displacement of the NW-trending faults, but also for the clear evidence that these strike-slip faults must pass upwards into low-angle thrusts.

Formerly, folding in the Borrowdale Group volcanics was taken to be evidence that the Iapetus closed, with associated compression, during the Ordovician Period (around 450 million years before the present). However, current thinking assigns this deformation to movements connected with the collapse of the main lava chamber of an ancient volcano.

SHAP FELL (NY 554057–556050)
F. Moseley

Highlights

The site provides an unrivalled continuous section through typical folds in the Upper Silurian rocks. It allows detailed observation of the style of these folds and, together with the Tebay site, demonstrates the geometry of the larger-wavelength folds in the Lake District.

Introduction

The Crookdale Crags section is situated alongside the A6, approximately 400 m south of Shap summit (Figures 3.16 and 3.17). It is entirely within the confines of the north-west limb of the Bannisdale Syncline, or Synclinorium, (a major F_1 fold). It exposes a sequence of the transition beds, which separate the Coniston Formation Grit from the Bannisdale Slate Formation (all Ludlow Series, Upper Silurian).

The general stratigraphy and structure of the Silurian rocks of Shap Fell and the surrounding country has long been known from the Old Series

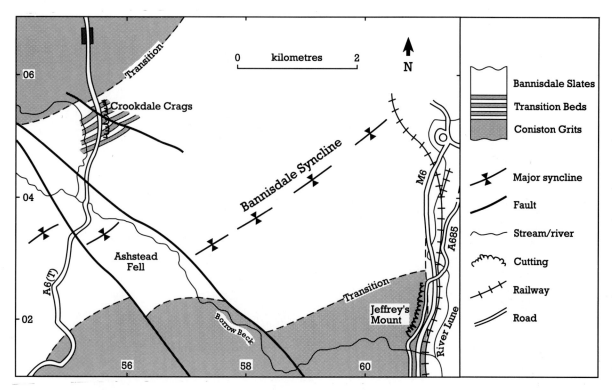

Figure 3.16 Geological map of the Bannisdale Syncline, showing positions of Crookdale Crags (see Fig. 3.17) and Jeffrey's Mount (after Moseley, 1986).

Geological Survey map. More recent descriptions of this and adjacent areas have been given by Moseley (1968), Soper and Moseley (1978), and Lawrence *et al.* (1986).

Together with the Jeffrey's Mount site, which lies on the south-east limb, the two sites provide cross-sections through the opposing limbs of the Bannisdale Syncline at very similar structural levels (Figure 3.16).

Description

The strata of this region are, in ascending stratigraphical order:

1. The Coniston Grit Formation, consisting of massively bedded greywacke with subsidiary, thin mudstone partings.
2. The transition beds, typified by the road section (Figure 3.17), consist of alternations of greywacke and mudstone. The former, in beds between 0.1 m and 2 m thick, display the same sedimentary features as the Coniston Grit Formation, whereas the latter range from less than 0.1 m thick to 7 m or more in thickness. There are also siltstones and fine greywackes.
3. The transition beds pass by gradation into the Bannisdale Slate Formation, with banded mudstone the dominant lithology. The banding generally results from alternations of pale silty laminae and thicker, dark mudstone bands.

These formations are disposed on the major (F_1) Bannisdale Syncline, or Synclinorium. This is an asymmetrical fold, with a longer and steeper north-west limb and a half wavelength exceeding 8 km. Numerous minor parasitic folds, with half wavelengths ranging from 2 m to 200 m, are located on the limbs of this fold (Figure 3.18). Generally, the massive, competent Coniston Grit Formation, on the extreme flanks of the syncline, is uncomplicated by minor folding, but the overlying transitional beds and the Bannisdale Slate Formation exhibit numerous minor folds with differing arrangements in different parts of the syncline.

Cleavage (S_1) is strongly developed in the mudstones, less strongly in the siltstones and scarcely at all in the massive greywackes. It is noteworthy that the axial planes of the folds are oblique to cleavage, with trends of 055° and 060° respectively. There are also dextral and sinistral wrench faults which are complementary to the folds.

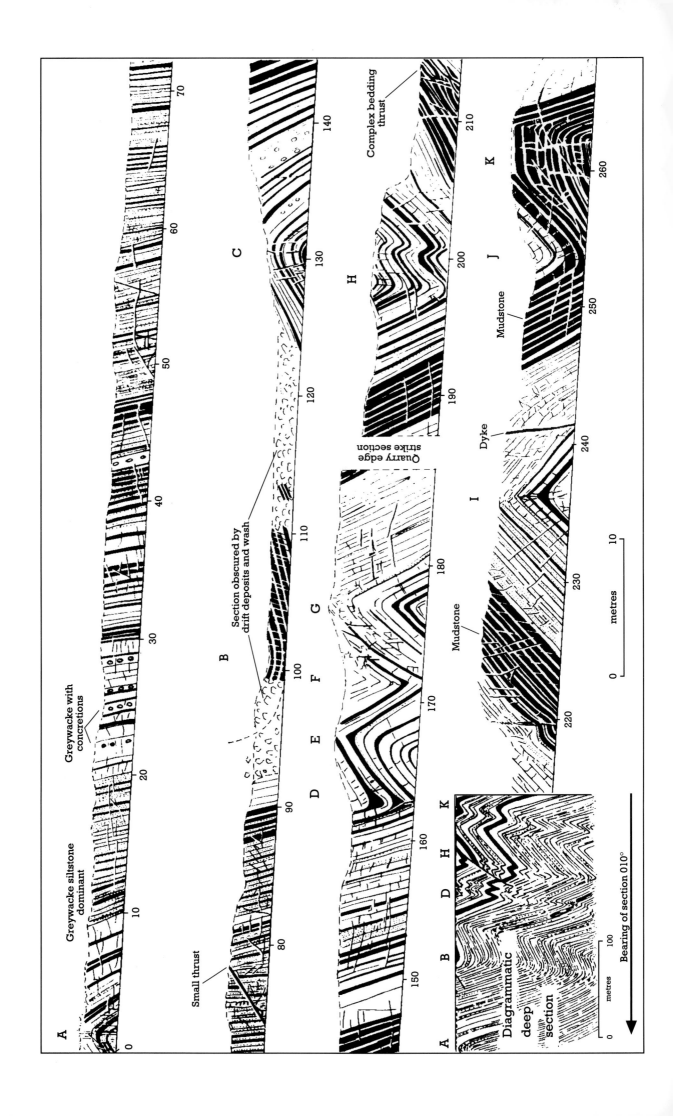

Greywacke siltstone dominant

Greywacke with concretions

Small thrust

A

Section obscured by drift deposits and wash

B

C

D

E

F

G

Quarry edge strike section

Complex bedding thrust

H

Mudstone

Mudstone

Dyke

I

J

K

Diagrammatic deep section

A   B   D   H   K

metres

0   100

Bearing of section 010°

metres

0   10

The A6 section illustrates the two styles typical of the north-west limb of the Bannisdale Syncline, that is, belts of sharp folds alternating with belts of uniform, steeply inclined strata (Figure 3.17). The section comprises 300 m of transitional beds and exposes nine folds, together with numerous associated structures such as cleavage with cleavage refraction, slickensiding, and disharmony.

Folds

The smaller folds within the fold belts are asymmetrical, with longer and more steeply inclined south-east dipping limbs (Figure 3.17). Fold plunge is constant, generally between 5° and 10°, to about 055°. Lineations related to the folding are found as slickensiding on the bedding planes. This is, apparently, bedding-plane slip formed during folding, but it is oblique to the direction of dip. However, if the folds are rotated so that the axes are horizontal, these slickensides become exactly downdip. Since the amount and direction of plunge is very similar to the dip of nearby Carboniferous rocks, the suggestion is that the fold plunge represents a post-Carboniferous tilt. Axial planes generally bisect the fold limbs, with beds having identical thicknesses on opposite limbs.

Varying competence of different lithologies during folding is revealed, particularly, by the differing behaviour of greywackes and mudstones where they are interbedded. The transitional beds between the Coniston Grit Formation and Bannisdale Slate Formation illustrate the effects best, with the sandstone bands showing near-concentric folding and the incompetent mudstones thickening out in axial regions with, concomitant development of strong cleavage (Figure 3.18). Measurements indicate (Moseley, 1968) the amount of axial thickening both in mudstones and in massive sandstones, which do in fact show this effect to a lesser extent. These measurements are given below as a ratio of the axial thickness (measured normal to the fold axis in the axial plane) to the limb thickness (measured normal to bedding):

1. Mudstone – 2.2:1 (mean of 13 mudstone bands, with an average limb thickness of 0.28 m);
2. Sandstone – 1.3:1 (mean of 18 greywacke bands, with an average limb thickness of 0.51 m).

Cleavage

Cleavage is strongly developed in the mudstones, less strongly in the siltstone and scarcely at all in the massive greywackes. It varies between genuine flow cleavage in some of the finer-grained mudstones, to much more common fracture cleavage. Unlike Silurian outcrops to the south and south-west, no second phase of cleavage has been detected. There is, nonetheless, some complexity and no straightforward relation to folding. For example, cleavage has different orientations on opposite fold limbs and it is rarely parallel to the axial planes of the folds. Of particular significance, is the clockwise angle of about 5° generally found between the strike of the cleavage and of the axial planes (and between cleavage and axial trend, since these structures are high angle and the plunge is low). It is therefore apparent that cleavage–bedding intersections are not indicators of fold plunge hereabouts, and field measurements clearly show this to be the case, particularly on south-east-dipping fold limbs.

One further complication is to be found in refraction of cleavage, usually sharply bent at mudstone–siltstone junctions, frequently curved in sympathy with graded bedding, but often bending into the sinistral joint set. This latter phenomenon of cleavage refraction occurs with downward passage from mudstone to greywacke.

Interpretation

The rocks at Crookdale Crags were deformed (folded, cleaved, and faulted), during the main Caledonian deformation (D_1), as a result of continent–continent collision. Interbedded sequences of sandstones and mudstones exhibit the best examples of fold structure in the Lake District and the transition beds at Crookdale Crags provide an excellent example of one such sequence. The style of deformation in these Silurian rocks is comparable with, and of the same age as, that exhibited in analogous Skiddaw Group lithologies, with the notable absence of complexities due to superimposed folding.

It is important to note that the cleavage is not

Figure 3.17 Fold structure along A6 road-cuttings at Crookdale Crags, Shap (after Moseley, 1968).

Figure 3.18 Shap Fell. D₁ folds developed in Silurian greywackes; cleavage can be seen in the interbedded muddy siltstone. View to east. (Photo: J. Treagus.)

axial planar to the folds. Cleavage transects the axial planes of the folds by approximately 5–10° in a clockwise direction. This is a common, but important, feature observed regionally in late-Caledonian structures of the Lake District and also in Wales (Woodcock *et al.*, 1988) and the Southern Uplands (Stringer and Treagus, 1980). It has been attributed to sinistrally oblique transpression – see Chapter 1.

This section exemplifies the typical structure of the north-west limb of the Bannisdale Syncline with its alternating belts of sharp folds and belts of steeply uniformly inclined strata and north-west vergence of minor folds. Soper and Moseley (1978) used this vergence to delineate the axial trace of the fold and noted that these belts are comparable, in width, to the zones of intense and weak cleavage in the Borrowdale Volcanic Group and suggested that they might be analogous structures.

This section should be compared with the structure at the Jeffrey's Mount site on the south-east limb of the Bannisdale Syncline (Synclinorium) where the folding is less intense.

Conclusions

This site, probably the most visited in the Lake District, provides an opportunity to examine the style of Caledonian deformation that characterizes the rocks of the Windermere Group. It is particularly important for showing details of this style, especially that folds are inclined in a south-easterly direction on the northern limb of a major syncline. The cleavage (closely spaced, parallel fractures) is not precisely parallel in plan view to the axes of the folds; this is an unusual relationship, although typical in the Silurian of the Lake District. This has been explained as a result of the way in which the Iapetus Ocean closed and the continents to the north (Laurentia) and south (Avalonia) came together and collided. It is thought that, at the time of this coming together, the margins of the northern ('North American') and southern ('European') continents were oblique to the direction of closure. Thus the intense, compressive stress imposed on the rocks by the collision has left graphic evidence, even demonstrating the orientation of the continental margins around 400 million years ago as Iapetus was closed.

**JEFFREY'S MOUNT, TEBAY
(NY 607017–610026)**
F. Moseley

Highlights

Clean, continuous roadside exposures here allow uninterrupted observations to be made on the D_1 folds in the Ludlow Series Coniston Grit Formation sandstones. These typify much of the structural style of the southern Lake District. Together with the Shap site, the locality permits a view of the larger-scale folds of the late-Caledonian deformation, produced during the Early Devonian, overprinted by numerous D_2 folds and fractures.

Introduction

This section is situated on the A685 about one mile south of Tebay village (Figures 3.16 and 3.19). It comprises the predominantly greywacke sequences of the Coniston Grit Formation. In this region, the Bannisdale Slate Formation and the Coniston Grit Formation were folded during the main Caledonian deformation (D_1) to form the major Bannisdale Syncline (F_1), and were affected by associated cleavage (S_1), minor folds (F_1) and faults.

The site shows many structural features similar to those seen at Shap Fell, but complements that site by its excellent exposures of structures on the opposing limb of the major, D_1 Bannisdale Syncline at a similar structural level. Further details on the general stratigraphy and structure of this area are given in the site description for Shap Fell. Mention of the site is made by Moseley (1972).

Description

At this section the Silurian Coniston Formation Grit consists of greywacke beds from 0.05 m–3 m thick, with subsidiary laminated greywackes, silt-stones and mudstones. There are nine minor folds present, all essentially concentric and with virtual zero plunge, but with complications where thicker mudstone units are involved. There are other complexities, which include small (~1 m wavelength) recumbent folds, low-angle shears, kink bands which deform the cleavage and high-angle

wrench faults (Moseley, 1972, 1986). Cleavage is poorly developed.

Folds

The folds and the poorly developed, axial-plane cleavage are the product of early Devonian main phase (D_1) deformation. Local D_2 phase structures (Table 3.1) according to Soper *et al.* (1987) post-date the underlying Shap Granite (394 Ma BP). The faults belong to the subsequent north-east-trending wrench fractures, some of which pre-date the nearby Carboniferous succession. The folds (see Figure 3.19) are open and, being on the south-east limb of the synclinorium, there is a general younging of the sequence towards the north-west and vergence to the south-east. The preponderance of greywacke sandstone in both thick and thin beds, as well as resulting in an absence of cleavage, also results in an absence of minor structures, although fold 4 exhibits a recumbent buckle on its northern limb. This could represent the late-Caledonian recumbent fold phase (F_2), which has been noted elsewhere in descriptions of the Silurian of the Lake District (Moseley, 1972), but it is more likely to be a local reaction to irregular stress conditions. The stereogram (Figure 3.19, inset) summarizes the orientations of bedding and fold plunge.

Faults

The most prominent faults are near-vertical wrench faults, with north and north-west trends, the former sinistral and the latter dextral. The actual displacements of the faults cannot be determined from the section. Both fault sets, however, intersect the roadside section where they have resulted in thin shatter zones (up to 1 m wide). There are also well-developed joints parallel to the faults. Small thrusts or 'thrust joints' are also fairly common (Figure 3.19), all inclined north, and generally with displacements of only a few centimetres.

Interpretation

The structures at Tebay are a product of main phase, early Devonian deformations (D_1). Comparing them with those on the north-west limb of the Bannisdale Syncline at Shap Fell, it is evident that the folding is less intense here. This is attributed to the competence of the greywackes at Tebay

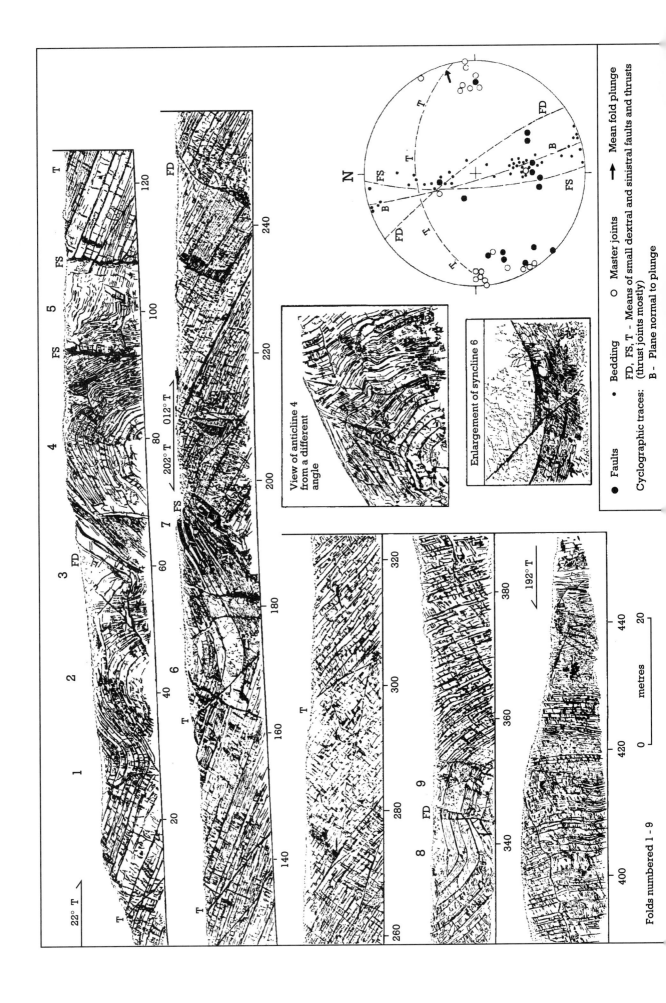

View of anticline 4 from a different angle

Enlargement of syncline 6

22° T

FD

FS 5

FS

FS

FD

T

T

3

4

1

2

20

40

60

80

100

120

202° T   012° T

FS 7

6

140

160

180

200

220

240

FD

T

T

260

280

300

320

8   FD   9

192° T

340

360

380

400

420

440

0   20

metres

Folds numbered 1 - 9

N

FS

T

B

FD

T

FD

T

FS

B

FD

T

T

● Faults

○ Master joints

• Bedding

FD, FS, T - Means of small dextral and sinistral faults and thrusts
(thrust joints mostly)

Cyclographic traces: B - Plane normal to plunge

➤ Mean fold plunge

compared with that of the mudstones of the A6 section, which also explains the poor cleavage development.

Since these exposures are at the same structural level as those at Shap Fell they provide an interesting contrast in fold style, in the context of variations in the multi-layer sequence and the development of cleavage. They also provide an opportunity to demonstrate the presence and geometry of a major D$_1$ fold, the Bannisdale Syncline, seen from the opposing vergence as well as from the calculation of sheet-dip and the demonstration of younging. Minor faults are particularly well seen here. Sections such as this, as well as that at Shap, have provided important data for the study of the late-Caledonian deformation and will do so in the future.

Conclusions

Jeffery's Mount provides an excellent cross-sectional view of the style of folding in the rocks of the Windermere Group (which are here of late Silurian age), when it is dominated by sandstones. It allows a comparison of the minor structures on the southeastern limb of one of the largest folds in the Lake District, the Bannisdale Syncline, as well as enabling the geometry and position of the major fold to be established. This major fold was the product of folding during the main Caledonian mountain building phase. Numerous later, flat folds, angular folds (kink bands), and low-angle fractures deform the main structures. These are known to be later than the granite intrusion at Shap, which is dated at 394 million years before the present, which gives an age to this final episode of Caledonian deformation.

Figure 3.19 Fold structure at Jeffrey's Mount, Tebay (after Moseley, 1972).

HELWITH BRIDGE (SD 803700)
J. E. Treagus

Highlights

The site provides continuous exposure across an anticline affecting Lower Palaeozoic rocks. This fold shows the ESE trend of the Caledonian structures, in some of the easternmost exposures in Britain. The exposures also exhibit the slight anticlockwise transection of the folds by cleavage. Both the ESE trend and the transection sense are considered to be important evidence for the geometry of the southern margin of Iapetus and of its movement during closure.

Introduction

Twenty kilometres south-east of the Lake District, the Lower Palaeozoic rocks outcrop as a 15 km^2 inlier surrounded by Carboniferous strata. The principal inlier, centred on Horton-in-Ribblesdale, between Austwick and Malham (see Figure 3.1), contains Arenig Series (Ingleton Group) and Ashgill Series (Coniston Limestone Formation) rocks, but is predominantly composed of later, Silurian turbidites and siltstones. The stratigraphy and structure of the inlier has recently been revised and reviewed by Arthurton *et al.* (1988), building on previous structural work by King and Wilcockson (1934).

The dominant structure is the ESE–WNW-trending Studrigg–Studfold Syncline, which preserves Ludlow Series rocks in its core, in the envelope of Wenlock formations. This fold is flanked by the Crummock Anticline to the north and the Austwick Anticline to the south, with a wavelength of some 3 km. Smaller-scale folds, with wavelengths of up to 300 m, open to upright style and plunges up to 25° to the ESE, are seen in most formations but particularly within the Horton Formation of Ludlow age (Arthurton *et al.*, 1988), partly equivalent to the Horton Flags of King and Wilcockson (1934) and the Horton Formation of McCabe (1972) and McCabe and Waugh (1983). The lithology is laminated, micaceous, somewhat calcareous, sandy siltstones, which generally exhibit a well-developed, spaced cleavage. The formation correlates with the Upper Coldwell Beds, below the Coniston Grit Formation of the Lake District. King and Wilcockson (1934, Plate 1 and Figure 7) mapped the folds in the area of the site, between the Arcow Wood and

The attitude and geometry of the fold are best illustrated by a profile section across the hinge region. The one illustrated was taken about two-thirds up the exposure (SD 80327006) and the data are presented here in a stereogram (Figure 3.20). The data in this and similar sections show the fold to be almost perfectly cylindrical, open in style, with axial plane (constructed) 119/88°N and plunge 18/118°.

The cleavage can be measured almost anywhere in the exposure and is approximately axial-planar, being subvertical and trending ESE. However, even by eye, an impression can be gained that the cleavage does, in fact, transect the hinge region in an anticlockwise sense. Precise measurement, across the hinge region reveals a very slight fanning, with very steep dips towards the south on the north limb, but subvertical on the south limb (Figure 3.20). More importantly, however, the non-axial planar relationship of the cleavage is confirmed by strikes up to 10° anticlockwise of the constructed axial plane and by more gentle intersection lineations (which are well developed on both cleavage and bedding) on the south limb (for example, 12/108°) than on the north limb (for example, 20/110°). Figure 3.20 clearly illustrates how the anticlockwise strike of the cleavage from the axial plane and fold hinge produces this relationship.

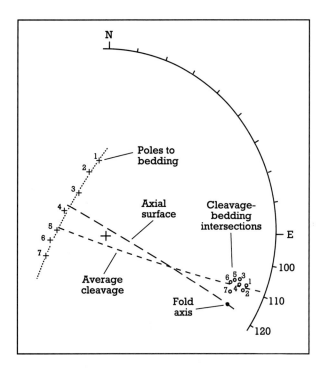

Figure 3.20 Stereographic representation of data from Helwith Bridge. Poles to bedding (crosses) numbered across the anticline with corresponding numbers at bedding–cleavage intersections (open circles).

Combs Quarries, showing the 100–300 m wavelength, the open style, axial trend to 120° and plunge up to 20° to the ESE.

The anticline that has been selected here is the anticline shown by King and Wilcockson (1934, Plate 1:SD 803700) south-east of the Foredale Lime Quarry. This fold is exceptionally well exposed in three dimensions, with the hinge plunging subparallel to the hillside.

Description

The principal exposure is in the hinge region of the anticline, which is seen almost completely in a 40 m-wide exposure on the hillside (SD 80357002–80277008). However, isolated exposures, to the south-east and east of the exposed hinge, give a more complete picture of the extreme limb dips, which reach 130/35°NE on the north limb and 087/35°S on the south limb. In the north of the site, bedding turns into the adjacent syncline.

Interpretation

The relationships described have been measured recently by Soper *et al.* (1987) in their analysis of the fold and cleavage swing across north-west England. The site was chosen to exemplify two features of the easternmost exposures available; firstly, that the swing in strike reaches ESE, secondly, that in contrast to the clockwise transection usually seen in the western and central Lake District, it is here distinctly anticlockwise.

The first of these features is clear from the bedding and cleavage data presented (Figure 3.20) and should be compared with that at Shap Fell and Tebay. This trend appears to be common to all the rocks in the inlier, according to the data of King and Wilcockson (1934) and Arthurton *et al.* (1988 and accompanying map sheet 60 of 1:50 000 BGS Series). The latter authors comment (p. 97) that the folds decrease in wavelength and amplitude to the north-west. The 20° plunge of the fold also appears to be typical of that in the inlier, although variation to 8° occurs both locally and regionally. In the Shap area, Moseley (1968) claims that the 5° easterly plunge is due to post-Carboniferous tilting. At Horton, the unconformable Carboniferous is virtually horizontal, so it would appear that the dominantly north-easterly plunge of the western Lake District is re-established here.

The anticlockwise transection of the axial planes by the cleavage is subtle, but measurable. The strike difference between the two subvertical planes is about 8° and produces plunge angle variations of up to 15° for intersection lineations. As explained in the Introduction to this chapter, a clockwise non-axial plane relationship has been increasingly recognized in the Southern Uplands, the Lake District and Wales. It is generally ascribed to be the result of sinistrally oblique compression which affected most of the British Caledonides during the final closure of Iapetus.

The reader is referred to Soper *et al.* (1987) for details, but the interest and importance of the Horton transection is that the change from a clockwise to this anticlockwise sense coincides with the swing in regional strike. This swing is interpreted as reflecting the detailed geometry of Caledonian structures on the southern flank of Iapetus. There they were moulded around the Precambrian Midland Massif during transpressive N–S closure of the ocean. On a regional scale this swing is related to the 'third arm' of the Caledonides, which runs from the North Sea into the German–Polish Caledonides (Tornquist's Sea Convergence Zone – see Figure 2, in Soper *et al.*, 1987). Thus, the change in transection sense described and discussed above is suggested to be a direct reflection of the contact strain around this Midland indenter and is therefore an important element in the understanding of both the geometry of the southern margin of Iapetus and its closure.

Conclusions

This site is important, not only as an example of the swing in Caledonian structure to ESE in the eastern Lake District, but it is also an example of the change in the east of the orientation of cleavage (fine, parallel fractures) relative to the ESE trend of the folds. This 'anticlockwise transection' of the folds by the cleavage is significant. Both features have been used recently to demonstrate the progressive swing in the trends of Caledonian structures around the Midland Platform. Soper *et al.* (1987) thought that the trend of the folds was modified by the solid mass of the English Midlands, against which the Lake District was forced when the latter collided with the northern (Scottish–American) continent as the Iapetus Ocean closed, in the final stages of mountain-building (orogeny).

JUMB QUARRY, KENTMERE (NY 449074)
A. M. Bell

Highlights

This is an important site, where volcanic accretionary lapilli have been the source material for significant quantitative analyses of the strain associated with the formation of slaty cleavage. It is the only site in the Lake District where such detailed strain analyses have been made.

Introduction

The quarry exposes part of the Wrengill Formation of the Borrowdale Group (Soper and Numan, 1974) which, here, is composed of water-worked air-fall tuffs of broadly andesitic composition (Figure 3.22A). Little work has been done on these volcaniclastic sediments (Moore and Peck, 1962; Bell, 1981), but particles within the tuffs have been used as indicators of strain since the work of Sharp (1849).

First estimates of strain using the deformed lapilli were made by Green in 1917, but the major period of quantitative strain analyses was started by Oertel (1970) who sampled at Jumb Quarry and concluded that the originally spherical lapilli had compacted during diagenesis and subsequently been distorted by strain during slaty cleavage formation. Publication of his results sparked off a heated debate in the literature between Oertel (1970, 1971, 1972) and other workers (Helm and Siddans, 1971; Mukhopadhyay, 1972; Ramsay and Wood, 1973; Wood, 1974) who argued that the lapilli shape indicated a period of cleavage-producing strain which had been superimposed on variably shaped, but undistorted lapilli.

At the very heart of this debate was the question of the origin of slaty cleavage; did the cleavage plane represent a principal plane of the total strain ellipsoid or was it the principal plane of some component of the total strain? Further work in

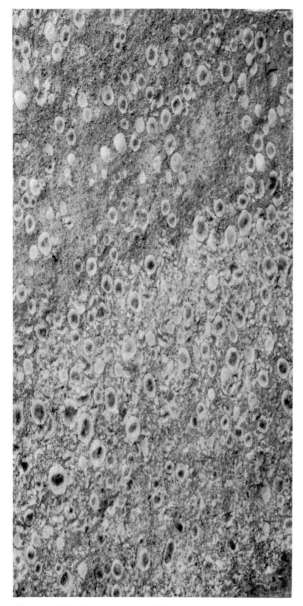

Figure 3.21 Jumb Quarry. The deformed accretionary spheres of volcanic ash have been used to measure the Caledonian strain in these Ordovician rocks. The cleavage plane photographed is 30 cm high and shows the intersection of bedding plunging to the left. (Photo: Dept of Geology, Manchester University.)

is related precisely to the shape of the tectonic strain ellipsoid.

Description

Jumb Quarry lies wholly within a tuff unit of the Wrengill Formation of the Borrowdale Group. It is one of a series of roofing-slate quarries in these beds in the Lake District, but here the lapilli horizons were presumably worked for their decorative qualities. The lapilli are contained within a distinctive tuff deposit, which has been traced as a stratigraphical marker horizon across the whole of the Borrowdale outcrop, and which has been invaluable for correlation within the upper part of the volcanic succession (Soper and Numan, 1974; Bell, 1981).

The tuff is composed of a sequence of light to dark green, fine- to coarse-grained, bedded air-fall tuffs, which are interbedded on a scale of several centimetres with structureless poorly-sorted tuffs containing lapilli. The lapilli themselves are distinctive ellipsoidal, or sub-ellipsoidal objects composed of a central core of coarse ash identical to the matrix, which grades systematically to an outer shell of light-green to white, fine ash (Figure 3.21).

Superimposed on the tuff sequence is a pervasive slaty cleavage (late Caledonian, S_1). The cleavage planes themselves are formed by the almost perfect alignment of minerals on the cleavage face. The cleavage planes dip steeply to the north-west (typically 80 degrees) and the mineral lineation is steeply plunging on that plane. Bedding dips variably from 30–70°SE, so the bedding–cleavage intersection lies almost at right-angles to the mineral lineation. The cleavage is not folded, except by infrequent minor kink bands.

Interpretation

Sedimentary structures and the absence of evidence for marine conditions, suggest that the tuffs in this area were deposited into shallow freshwater lakes, lying on the flanks of the major volcanic centres. The nature of the lapilli indicates they are fossil 'hailstones' and were most probably the products of a violent eruption which produced a massive ash- and steam-laden thundercloud above a volcanic crater. They would have oscillated within this cloud until their increasing weight, or the waning of the energy of the eruption, allowed them to fall into the soft, wet, ashy mud.

other slate belts (Siddans, 1972; Wood, 1974) and a more extensive survey and analysis of the Borrowdale Group lapilli (Bell, 1981, 1985) has produced a solution which suggests that the observed particle shape is the product of strain modification, both diagenetic and tectonic, of variably shaped lapilli, but that cleavage formation

The ellipsoidal lapilli make excellent strain markers, since their composition, and hence their competency, does not differ from that of their matrix. They are also found in large enough numbers to give statistically valid results. Each lapillus is now roughly ellipsoidal (some exactly so) and at Jumb Quarry the short axis of most lies roughly normal to the cleavage plane, and their long axes lie roughly along the mineral lineation on the cleavage face. They show that there can be no doubt that compression across the cleavage produced the planar fabric and an accompanying extension within the plane of cleavage caused the mineral lineation.

Estimates of 66% shortening, normal to cleavage, were made by Green in 1917, the first documented strain analysis using the accretionary lapilli. Oertel (1970) calculated a single ellipsoidal shape which was an average of many measured lapilli. He then factorized this into a plane strain (conserving volume and with no change of length along the intermediate axis) oriented in the cleavage frame, that is one ellipsoid axis normal to the cleavage plane, one within the cleavage plane parallel to the mineral lineation and an oblate strain (long and intermediate axes equal) oriented in the pre-cleavage bedding frame. He concluded that the tuffs had compacted by some 50% normal to bedding prior to cleavage formation, and then shortened by about 50% across the cleavage plane. These results were broadly in agreement with Green's estimate.

Helm and Siddans (1971) repeated the analysis using Rf/ϕ technique (a graphical technique which allows the calculation of the finite strain (Rf) from measurements of a population of elliptical strain markers which show a variation of orientation ((ϕ) of their axes) on specimens from the nearby Steel Rigg Quarry and elsewhere. Their results seemed to indicate that elliptical lapilli were randomly oriented prior to cleavage formation. Their cleavage strain was not plane ($k = 1$), but had a k value of 0.4, and the mean-shape of the lapilli differed from the cleavage frame by less than five degrees.

The lengthy debate that followed these studies, concerned the origin of slaty cleavage and focused on the relationship of the strain ellipsoid to the cleavage plane. It was continued by Bell (1981) who analysed samples from nine localities including Jumb Quarry and Steel Rigg Quarry using both an Rf/ϕ technique and an average ellipsoid factorization technique similar to that used by Oertel. His results (Figure 3.22A and B) showed

that, in localities of intense cleavage and steep bedding like Jumb and Steel Rigg, the lapilli had probably been compacted by 66% in bedding prior to a tectonic strain which caused shortening of some 50–70% across cleavage and which was close to, but not exactly plane (k values from 0.8 to 1.2). Elsewhere along strike, particularly in localities that lay on the gentle north-dipping limb of a major fold, tectonic strains were much lower and the long axis of the average lapillus pitched at a high angle to the cleavage mineral lineation.

Subsequent work by Bell (1985) related the variation of tectonic strain along the outcrop, to the process of cleavage formation, concluding that the lapilli shapes supported a cleavage-forming mechanism that began as layer-parallel shortening, accomplished largely by volume loss, and developed into plane strain with conservation of volume with the onset of lowest-grade metamorphism and full cleavage development.

The modern consensus is that the strain which formed the cleavage is probably somewhat oblate ($k < 1$), rather than plane, and almost certainly has the cleavage plane as its principal plane. The lapilli have suffered another distortion as well as cleavage formation and this is most likely to have been compaction during diagenesis. Volume loss normal to the cleavage plane seems to be important in the earliest stages of tectonic deformation, even after diagenetic compaction.

Jumb Quarry provides important exposures of intensely cleaved volcanic accretionary lapilli within the Borrowdale Group, which have been used as strain markers. Opportunities for such accurate studies of strain are extremely rare in deformed rocks. They provide invaluable evidence for the calculation of crustal shortening. Results indicate that the late-Caledonian slaty cleavage (S_1) formed by compression and volume loss normal to the cleavage plane, and subvertical stretching. Strain measurement from this site suggests a 50–70% crustal shortening for this part of the Caledonides.

Conclusions

This locality makes it possible to actually measure the amount of compression which the Lake District suffered when continental collision brought about the closure of the Iapetus Ocean around 400 million years before the present. At Jumb Quarry, beds of volcanic ash contain accretions that were once near spherical and which are now near-perfect ellipsoids. In common with other rocks in

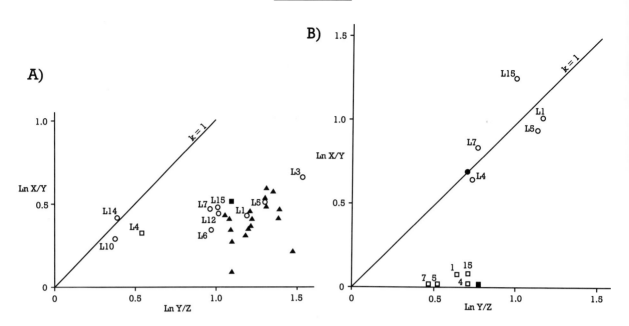

Figures 3.22A and B Flinn plots of average lapilli shapes (A) and actual strain ellipsoids (B) for accretionary lapilli horizons within the Borrowdale Volcanic Group. Ellipsoid long, intermediate, and short axes are denoted by X, Y, and Z respectively. (A) shows the range of overall lapilli shapes throughout the Borrowdale Group (data from Bell (1981 – open circles), Oertel (1971 – open squares), Green (1917 – solid squares) and Helm and Siddans (1972 – solid triangles)). (B) Bell (1981) resolved compaction strains (squares) and tectonic strains (circles). Compaction strains are uniaxial ((X = Y) > Z, k tends to zero) whereas tectonic strains are almost plane (k = 1) (data from Bell, 1981).

the volcanic Borrowdale Group, these lapilli were compressed and deformed by this mountain-building event, which geologists call the Caledonian Orogeny. This distortion of the originally spherical lapilli tells us that the rocks in this part of the Lake District were shortened by between 50% and 70% by extreme compressive strain caused by the collision of northern (Scottish–American) and southern (Lake District–European) continents.

Chapter 4

Wales

INTRODUCTION – A STRUCTURAL PERSPECTIVE

W. R. Fitches

Most of Wales is underlain by a thick pile of sedimentary rocks, ranging in age from Cambrian to Devonian, together with volcanic and intrusive rocks which are exposed mainly in the north and south of the country. These rocks collectively constitute the fill of the Welsh Lower Palaeozoic Basin, a major region of prolonged subsidence which developed as a marginal basin on the southern continental flank of the Late Precambrian–Early Palaeozoic Iapetus Ocean. It was deformed and weakly metamorphosed, mainly in Early Devonian times, during the Caledonian Orogeny in response to plate collision processes.

The edges of the basin, at least, are floored by a Precambrian–earliest Cambrian basement, which is exposed in Anglesey, the northernmost Welsh mainland, the Welsh Borders, and in South Wales. Devonian, Carboniferous, and younger strata hide the eastern part of the basin, although the transition with the Midland Platform of England is exposed within the Welsh Border Fault System and the classic Shropshire sections. In South Wales, basin and platform sequences are largely buried by Devonian and Carboniferous rocks and, with them, are caught up in the northern part of the Late Carboniferous Variscan orogenic belt.

The Welsh Lower Palaeozoic Basin is one of the world's classic geological regions in the sense that many of the basic principles of geology were first formulated there in the last century and early part of this century. The earliest work in Wales, concerned with structural geology, is discussed and referenced by Bassett (1969) in his comprehensive essay on Early Palaeozoic major structures and requires no further comment here. The observations and interpretations made by O. T. Jones, R. M. Shackleton and others earlier this century, however, remain pertinent to modern research as the following pages reveal. Since Bassett's review, research into the tectonic evolution of the Welsh Basin has undergone a renaissance, particularly in the last decade, as new ideas about sedimentary basin dynamics and deformation have been applied.

To provide a context for the selected Geological Conservation Review sites in Wales, this chapter concentrates mainly on descriptions of the main structural characteristics of the Lower Palaeozoic succession in Wales. However, to appreciate fully the importance of these sites to the evolution of ideas and to ongoing research, it is also necessary to comment briefly on the tectonic models which have been applied to Wales. It is no longer realistic to explain the deformation patterns in terms of a single and simple end-Caledonian event; the basin was tectonically active throughout its Lower Palaeozoic extensional, subsidence history as well as during end-Caledonian compression. Most of the structural characteristics of the Welsh Basin can ultimately be attributed to reactivation of basement structures.

Folds

Folds in the Welsh Basin range in magnitude from those which can be identified on small-scale maps such as the 1:625 000 Geological Survey Ten Mile Map of Great Britain (BGS 1979) and the 1:1 584 000 Tectonic Map of Great Britain and Northern Ireland (BGS 1966), to those which are visible only under the microscope. The axial traces of the major folds are shown in Figure 4.1.

The largest folds have wavelengths of several tens of kilometres and include the Snowdonia Synclinorium and Harlech Dome of North Wales, the Towy Anticline of south-east Wales and the Plynlimon Dome, central Wales Syncline, and the Berwyn Dome of mid-Wales. These structures are usually periclinal and thus commonly produce broadly ovoid outcrop patterns of rock units. Several of these largest-scale structures are not necessarily simple products of end-Caledonian horizontal shortening. Some of the domes may represent horsts which rose intermittently (or failed to subside) during the extensional stages of basin development, buoyed up perhaps by low-density basement or Lower Palaeozoic acid intrusions. Examples are the Derwen Anticline and Harlech Dome in the north (Fitches and Campbell, 1987; Kokelaar, 1988) and the Towy Anticline in the east (for example, Tyler and Woodcock, 1987). Similarly, the Snowdonia Synclinorium is probably nucleated on a major Ordovician caldera (Howells *et al.*, 1986).

Major folds have wavelengths of 1–5 km and amplitudes of 1–2 km. Examples are the Idwal and Hebog Synclines, and Tryfan and Capel Curig Anticlines (Wilkinson, 1988), components of the regional Snowdonia Synclinorium (Figure 4.2), the Llangollen Syncline of north-east Wales (Wedd *et al.*, 1927), and the Capel Cynon Anticline of West Wales (Anketell, 1987).

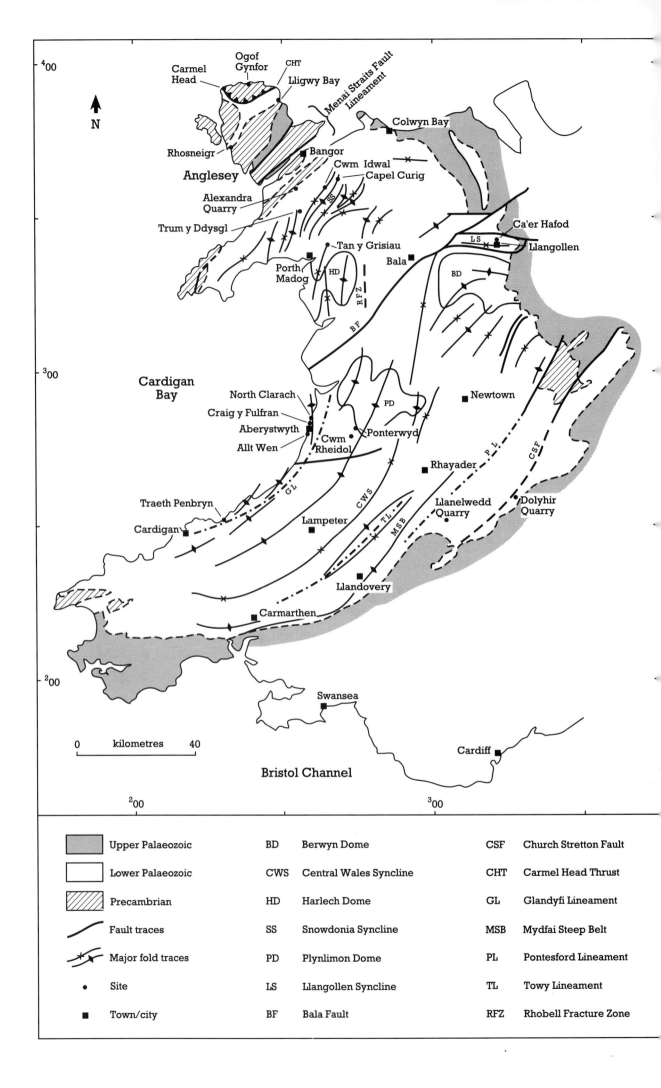

▨ Upper Palaeozoic	**BD** Berwyn Dome	**CSF** Church Stretton Fault
☐ Lower Palaeozoic	**CWS** Central Wales Syncline	**CHT** Carmel Head Thrust
▨ Precambrian	**HD** Harlech Dome	**GL** Glandyfi Lineament
⌒ Fault traces	**SS** Snowdonia Syncline	**MSB** Mydfai Steep Belt
⤳ Major fold traces	**PD** Plynlimon Dome	**PL** Pontesford Lineament
• Site	**LS** Llangollen Syncline	**TL** Towy Lineament
■ Town/city	**BF** Bala Fault	**RFZ** Rhobell Fracture Zone

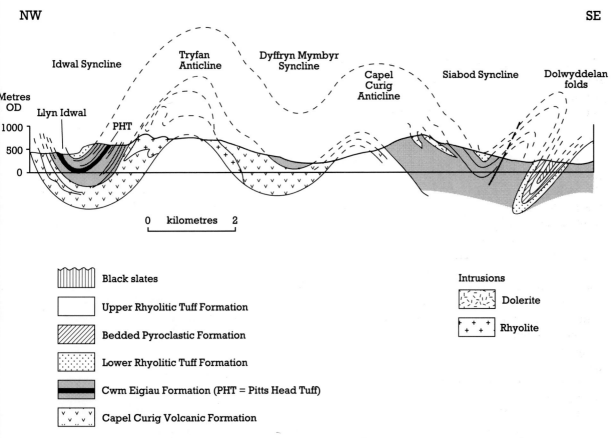

Figure 4.2 Section through the major folds of Snowdonia (after Wilkinson, 1988).

The Idwal Syncline is one of Britain's best-known structures and typifies these major folds. It is an open, almost symmetrical fold with upright NE–SW axial plane and gentle north-east plunge in Cwm Idwal. As with other major folds, however, it is non-cylindrical, non-plane, and its geometry changes markedly along its axial trace; to the north-east of the A5, in the slopes of the Carneddau, it becomes almost isoclinal and plunges steeply to the south-west. Less typically, the major folds are asymmetrical (and then usually close to tight) with one limb steep to overturned. Trum y Ddysgl is situated on the north-west overturned limb of the Idwal–Snowdon Syncline. The Hebog Syncline, the southern continuation of the same structure, also takes this form locally, where its axial plane shallows to about 60°NW from its usually vertical attitude, but the Dolwyddelan Syncline is reclined to strongly inclined along its entire length. Some major folds have an *en échelon* geometry, for example the Cwm Pennant Anticline described by Roberts (1979), Wilkinson and Smith (1988), and Smith (1988).

Smaller-scale folds, with wavelengths between 200 and 1500 m, are locally visible in hillsides and coastal cliffs, as on the coastline south of Aberystwyth (Price, 1962). Most are revealed only by detailed mapping, as in west Wales by Craig (1985, 1987) and Anketell (1987). Outcrop-scale folds, with amplitudes between a few centimetres and 10 m, are common in some areas (for example, Aberystwyth and its hinterland) but rare in others (much of Snowdonia, for instance). Because they mimic the morphology and orientation of the major folds, these small structures provide a ready source of information about fold geometries. Such folds are illustrated in Silurian rocks at Cwm Rheidol and in the Ordovician of Anglesey, for instance at Rhosneigr.

Figure 4.1 Map showing the traces of the principal folds and faults of Caledonian age in Wales. The localities described in the text are also shown.

Cleavage

Almost all rocks in the Welsh Basin are cleaved to some extent. Exceptions are the more rigid volcanic rocks, such as the welded tuffs of Snowdonia, and most of the major igneous intrusions such as the Tan y Grisiau microgranite of North Wales.

Cleavage is most clearly seen in argillaceous rocks, where it is usually a spaced (up to 5 mm), disjunctive fabric (using Powell's (1979) classification). Cleavage domains, the sites of pressure solution and growth of new phyllosilicates (Craig *et al.*, 1982), anastomose about detrital grains and diagenetic pyrite. The cleavage is more widely spaced and ill-defined in siltstones, sandstones, and volcaniclastic rocks. Several of the sites described below illustrate these cleavage character-istics. In mudstones and shales containing a bedding-parallel compaction fabric, the cleavage is commonly a zonal crenulation type. Exceptionally, the cleavage in mudrocks is continuous, giving rocks which yield high-quality roofing slates as in the Slate Belt of North Wales, especially well seen at Moel Tryfan. The Welsh slates inspired much early work concerning the nature of cleavage (for example, Sharp, 1849; Sorby, 1853), and more recently on cleavage-forming mechanisms (see Wood and Oertel, 1980; Whalley, 1973; Knipe and White, 1977; White and Knipe, 1978).

The margins of minor intrusions, notably those of dolerite dykes in Snowdonia, are commonly cleaved, the fabric being defined by aligned deformed vesicles and new metamorphic minerals such as actinolite and chlorite. Some of the plutonic bodies, including the Mynydd Mawr Granite in North Wales, also have cleaved margins.

On small-scale maps, such as Figure 4.1, the trace of cleavage usually appears to be parallel to fold axial traces. This congruence is because cleavage is approximately axial-planar to, or has a fanning relationship with, most folds. However, recent studies have revealed many examples of cleavage transecting folds. Craig (1987) showed that in parts of West Wales the cleavage is anticlockwise with respect to small folds, axial planar to larger folds, and clockwise to major folds. Cave and Hains (1986) reported numerous examples of clockwise transection from the Aberystwyth area and, in earlier unpublished work, were the first to record this phenomenon in Wales. The transection of folds by cleavage is widely recognized in the Caledonides of South Britain, where it has been interpreted as the consequence of transpressional deformation produced by oblique collision (for example, Soper *et al.*, 1987; Woodcock *et al.*, 1988).

Other complex fold–cleavage relationships are to be found in southern Snowdonia where, in the outer arcs of some folds, cleavage refracts so strongly that it becomes virtually parallel with bedding (Smith, 1988). Locally, as on the coast north of Aberystwyth, cleavage appears to have various time relationships to folding: it cuts across some folds, is axial planar to others, or is itself folded (Fitches and Johnson, 1978).

Strain

Strain markers are widely, although patchily, distributed throughout the Welsh Basin. Until recently, few attempts have been made to use them for strain analysis, despite the recognition of their potential use, over a century ago, by Sorby (1853). The now classical work on strain analysis in Wales, which had international significance, was carried out by Wood (1971, 1974), mainly in the Cambrian rocks of the Slate Belt in North Wales. He used as strain markers the centimetric, green reduction spots found widely in the grey and purple slates of that region. These spots, of diagenetic origin, were originally nearly spherical, but have been deformed into triaxial ellipsoids which have short axes normal to cleavage and long axes down the dip of that fabric. Wood calculated that up to 67% horizontal shortening and up to 157% vertical extension took place during cleavage development in the Slate Belt.

Siddans (1971) used accretionary lapilli to measure strains around the Capel Curig (or Mymbyr) periclinal anticline. Roberts and Siddans (1971) studied the shapes of various volcanic fragments in the Llwyd Mawr ignimbrite of Snowdonia, attempting to separate the volcanic compaction and tectonic strains. The strain data, obtained from some 50 sites by various workers throughout North Wales, were used by Coward and Siddans (1979) to calculate that the part of Anglesey between the Carmel Head Thrust and the Menai Straits has been shortened by 12 km during the end-Caledonian deformation, while the section from the Menai Straits to the Welsh Borders has been shortened by 43 km.

More recently, Wilkinson (1987, 1988) has made a major strain study of central Snowdonia, obtaining data from 250 sites by using a variety of strain markers, mainly in Caradoc volcaniclastic units:

volcanic clasts, accretionary lapilli and siliceous concretions. Similarly, Smith (1988), working in southern Snowdonia, has used as markers ferruginous ooliths and, in the aureole of the Tan y Grisiau Granite, contact metamorphic spots. Further south, Craig (1985) measured the low strains in parts of west Wales from deformed concretions, following on studies made by Lisle (1977) on the Aberystwyth Grits.

The more recent strain studies have revealed that, although deformation in the Welsh Basin almost invariably involves nearly horizontal shortening and vertical extension, the strain magnitudes and shapes of finite strain ellipsoids are highly variable, even within small areas. Wilkinson (1987, 1988), Smith (1988), and Wilkinson and Smith (1988) attributed this heterogeneity partly to variations in lithology and positions of sites in major folds. However, the main cause, in their view, is the location with respect to reactivated basement fracture zones; anomalously high strains are found in cover rocks above basement fractures, whereas lower and more homogeneous strains characterize the cover to blocks between fractures. This topic is further discussed below.

Faults

As information has accrued on sedimentation and volcanism in the Welsh Basin, it has become increasingly apparent that the basin was intermittently tectonically active throughout its extensional history; subsidence and sediment accumulation were accomplished to a large extent by faulting. According to some authors, notably Wilkinson and Smith (1988), these basin faults were nucleated on reactivated basement faults. Many faults were repeatedly active, particularly those comprising the Menai Straits (Gibbons, 1983) and Welsh Borders (Woodcock, 1984a, 1984b) lineaments. Most of these early faults have a Caledonoid (NE–SW) or nearly N–S strike (Fitches and Campbell, 1987).

Fault-control on sedimentary facies and thicknesses in the Lower Palaeozoic succession is now widely documented, for example: in the Ordovician of Anglesey (Bates, 1974); along the Bala Fault lineament (Fitches and Campbell, 1987) and in the Llandovery area (Woodcock, 1987a). Most of the syndepositional faulting is consistent with an extensional or transtensional regime. Similarly, the accumulation of volcaniclastic deposits and the location of intrusions (notably plugs and dykes)

were governed to a large extent by contemporary faulting. This control has been closely studied in central Snowdonia by the British Geological Survey's Snowdonia Unit, which has demarcated the Ordovician Snowdon caldera fractures and apical graben (Reedman *et al.*, 1985), for example, and identified a failed rift system from dyke distributions (Campbell *et al.*, 1988), while Orton (1988) has analysed the interaction between sedimentation and faulting in central Snowdonia.

Woodcock (1984b, 1988; see Llanelwedd and Dolyhir Quarries) has shown that several lineaments in the Welsh Borders comprise complex systems of dip-slip and strike-slip faults, with folds in places, which can be interpreted as strike-slip 'duplexes'. Woodcock (1984b) and Lynas (1988) described the Clun Forest Disturbance, a NNE–SSW linear zone of anastomosing faults and folds in the Welsh Borders, as a flower structure. This structure comprises upward divergent reverse faults in a linear zone of uplift, and was caused by strike-slip displacement. The Glandyfi vergence divide in west Wales (Cave, 1984; Cave and Hains, 1986) has been interpreted as another example of a flower structure by Craig (1985, 1987), who takes this to be an end-Caledonian structure probably nucleated on a long-lived basement fracture.

Thrust faults occur on a small scale in various parts of the basin (Price, 1958, 1962). The only large-scale example, however, is the Carmel Head Thrust on Anglesey, which carries rocks of the Mona Complex over Ordovician strata (Greenly, 1919; Bates and Davies 1981). The Tremadoc 'Thrust Zone', identified by Fearnsides (1910) and Fearnsides and Davies (1944) on the basis of repetition of strata and various small-scale structures, is reinterpreted by Smith (1987, 1988) as a Caradoc olistostrome which probably slid northwards off the Harlech Dome. In mid-Wales, Jones and Pugh (1915) identified the NNE-striking, west-dipping Brwyno–Gelli Goch and Cascade–Forge Thrusts; the throw on these structures is not known (Cave and Hains, 1986).

Much of Central Wales is cut by ENE faults which displace folds and other faults, including the mid-Wales thrusts, and hence are late structures. Those mapped by Cave and Hains (1986) have dominantly dip-slip displacement. Faults with this trend are mineralized in places (Phillips, 1972; Raybould, 1976) and are perhaps manifestations of early Variscan, rather than end-Caledonian deformation (Fitches, 1987). The Bala Fault, one of the major faults of the Welsh Basin (Figure 4.1), is considered by Fitches and Campbell (1987) to have moved by strike-slip displacement also mainly during Variscan events, although parts of that structure were active as dip-slip faults during the early Palaeozoic.

Bedding planes in many parts of the Welsh Basin have been used as detachment surfaces, which are usually marked by striated thin veins of quartz, carbonate, and other minerals. Particularly outstanding examples, which first attracted the attention of researchers (for example, Nettle, 1964 and Nicholson, 1966, 1978), are found near Llangollen (Ca'er-hafod). Others, with genetically associated vein breccias, bedding-normal veins and folds, are seen at various places along the west Wales coast (Fitches *et al.*, 1986). Nicholson (1966) showed that the Llangollen veins preceded end-Caledonian deformation, although Davies and Cave (1976) ascribed those in west Wales to pre-lithification gravity sliding which accompanied fold and cleavage development. More recently, Craig (1985) and Fitches *et al.* (1986) interpreted the association of veining, bedding-plane detach-ment, small-scale thrust and normal faulting as the product of post-lithification hydraulic jacking followed by gravity sliding before the end-Caledonian deformation. These intriguing structures remain of topical research interest – see Traeth Penbryn and Allt Wen.

Deformation sequence

Most parts of the Welsh Basin are characterized by a single set of folds on upright, mainly NE–SW axial planes which are accompanied by cleavage. These structures are commonly described as the 'regional' or 'main' structures. They are usually attributed to end-Caledonian or Late Silurian–Early Devonian deformation (Dewey, 1969), although the current trend is to use the North American term 'Acadian' (for instance, Soper *et al.*, 1987). Woodcock (1987a) supported Jones' (1955) view that the main cleavage-forming event, in south-east Wales at least, continued into, or was confined within, late early Devonian to mid-Devonian times. Ongoing isotopic work in central and North Wales, by J. Evans (BGS), M. Dodson and P. Bishop (Leeds University), is likely to shed light on the timing of deformation in those parts of the basin, and on whether or not the 'main' structures are contemp-oraneous across the basin. Particularly relevant in providing time constraints on the deformation is the Lligwy Bay section on Anglesey (Greenly, 1919; Bates and Davies, 1981) where presumed Devonian

red-beds are folded, cleaved and thrust: usually, elsewhere in Britain south of the Southern Uplands Fault Devonian rocks were only strongly deformed by Variscan events.

It has become clear that deformation of the basin cannot be regarded simply as a single, climactic event; several regions were tectonically active at various times during basin evolution, and even the main, end-Caledonian deformation was polyphase.

Along the N–S Rhobell Fracture Zone on the eastern flank of the Harlech Dome, Kokelaar (1977, 1979) demonstrated an early Tremadoc set of folds with steep N–S axial planes, which preceded eruption of the Rhobell Fawr Volcanic Group and, similarly oriented, end-Tremadoc folds which deform those volcanics. An end-Tremadoc regional deformation event has been recognized over much of north-west Wales on the evidence that the basal Arenig deposits are usually unconformable on older rocks (Shackleton, 1953, 1954; George, 1961; Roberts, 1979; Allen and Jackson, 1985a, 1985b). The magnitude of the sub-Arenig unconformity increases, irregularly, to the north and west of the Harlech Dome until some 5000 m of mainly Cambrian strata are cut out on Llŷn. Roberts (1979) suggested that this tectonic event involved block uplift on major faults. It is not yet clear whether this end-Tremadoc event caused widespread folding and cleavage or pro-duced these compressional structures only along fault zones.

Allen and Jackson (1985a, 1985b) tentatively attributed localized NW-trending folds in the Harlech Dome to a late Ordovician–Silurian deformation, but structures of this age within the basin are not widespread. There is evidence, however, (Woodcock, 1984b) for strike-slip faulting along the south-east margin of the basin, of probable Ashgill age. Lynas (1970a) considered evidence for a mid-Caradoc event producing a flat-lying cleavage around the northern and eastern flanks of the Harlech Dome, perhaps in response to incipient uplift of the dome associated with volcanism. Coward and Siddans (1979) attributed this fabric to deformation in the Tremadoc Thrust zone. However, this 'early' cleavage is now inter-preted as the regional, end-Caledonian cleavage, which has an unusually shallow dip because host rocks were ramped southward over the Tan y Grisiau Granite and Harlech Dome during the regional deformation (Bromley, 1971; Campbell *et al.*, 1985; Smith, 1987, 1988). The lineation on the low-angle cleavage, which earlier workers had ascribed to intersection with the upright, regional cleavage, is usually a grain shape fabric in the cleavage, generally the effect of elongate grains in the cleavage, probably caused by strong down-dip extension.

Several sets of end-Caledonian structures have been identified in various parts of Wales. Roberts (1967, 1979) recorded three deformations in North Wales. According to him, the main fold architecture, represented by the mainly NE–SW upright cleavage and folds such as the Idwal Syncline, developed first, during a D_1 event. Rare, recumbent folds with axial-planar crenulation cleavage were then imposed (D_2), and followed in turn by upright NW–SE to N–S folds and crenulation cleavage (D_3). Helm *et al.* (1963) considered that the D_3 deformation was responsible for the major arcuation of the main structures through North Wales. However, Roberts (1979) assigned the late structures to localized deformation (along zones above basement fractures according to Wilkinson, 1988); Coward and Siddans (1979) argued against this refolding model on the grounds of fold geometry.

Polyphase deformation structures are also known in mid-Wales. Fitches (1972) described, from the Aberystwyth area, a sequence of structures, closely similar to that described by Roberts (1969) in North Wales. Other examples from northern Mid-Wales were described by Martin *et al.* (1981). Tremlett (1982) and Craig (1985) suggested that the D_2 and D_3 structures probably denote localized movement near faults, rather than regionally correlatable events. Their view is supported by the close spatial association of kinks and crenulations of the regional cleavage with the Tal y Llŷn section of the Bala Fault system (Bracegirdle, 1974; Fitches and Campbell, 1987). On these grounds, the structures deforming the regional cleavage in Mid-Wales, and perhaps also Snowdonia, are Variscan rather than Caledonian – see above. In summary, the main D_1 deformation occurred in the late Silurian to early Devonian but not necessarily as a single climactic event. Earlier movements, often related to faulting and volcanic activity, caused local folding and tilting, but there is no evidence of earlier cleavage. Similarly, deformation later than D_1 is of local development and it may often be attributed to fault movement which may be Variscan.

Tectonic models

It has been shown that the main end-Caledonian folds throughout the Welsh Basin, are typically upright and non-cylindrical structures, irrespective of scale; and the main cleavage usually has an axial planar, fanning or transecting relationship with the folds. Despite this uniformity of morphology, however, the orientations of these structures are highly variable (Figure 4.1). Particularly conspicuous in North Wales is the arcuation of axial planes and cleavage from N–S in the Harlech Dome and southern Snowdonia, through NE–SW (Caledonoid) in central Snowdonia, to E–W in north-east Wales. A similar, but smaller scale, arcuation occurs about the north-western flank of the Berwyn Dome, where the NNE–SSW Central Wales Syncline turns abruptly into the E–W Llangollen Syncline. In much of central and west Wales the structures strike approximately NNE–SSW, but in the southern part of the basin they arc to ENE–WSW and then E–W. The cause of these arcuations has been extensively debated in the past and, in conjunction with the more recently recognized allied enigma of transecting cleavages, is again a topical research subject. An interrelated problem is whether or not deformation is 'thin-skinned', whether the structures flatten downward to merge with one or more detachment zones deep in the cover, or upper part of the basement, or is 'thick-skinned', that is the cover structures link with, and are partly controlled by, those deep in the basement.

Jones (1912) was the first to address the problem of arcuation in the southern part of the basin, concluding that it is a primary Caledonian feature and not a product of deflection of Caledonoid structures by a younger event. Anketell (1987) supported this view and contradicted Pringle and George's (1948) explanation which postulated subsequent warping. Shackleton (1954), concerned with northern Wales, considered that cover structures were strongly guided by fracture systems in the basement. He pointed out that on Anglesey, for example, at Ogof Gynfor, there is no detachment along the exposed cover–basement boundary and that there are small-scale examples of faults in the basement passing up into faults and folds in the cover (earlier recorded by Greenly, 1919). These and other observations led him to suggest that the structural arcuations in the cover in northern Wales are responses to moulding of structures against and above the sides and corners of basement fault-blocks.

By contrast, Helm *et al.* (1963) explained the

North Wales arcuation in terms of regional refolding of Caledonoid structures about later NW–SE axial planes. Their interpretation was countered by Coward and Siddans (1979) on the grounds that there are no structures in the region which are consistent with the inner-arc compression and outer-arc extension required by the refolding model.

Coward and Siddans (1979) went on to erect a 'thin-skinned' interpretation of northern Wales, based largely on their strain study, deducing that the cover and upper part of the basement lie on a deep detachment. They explained the structural arcuation as the result of compression of Snowdonia against a rigid block situated beneath the Berwyn Dome. Campbell *et al.* (1985) modified this indenter model, considering that the rigid block comprised the Caradoc Tan y Grisiau granite and basement rocks beneath the northern flank of the Harlech Dome.

In a radical departure from previous interpretations of deformation in Wales, Woodcock (1984a, 1984b) introduced the concept of strike-slip and transpressional tectonics to explain the end-Caledonian, and perhaps earlier, tectonic events in the south-eastern part of the basin. These views, together with an assessment of possible Variscan and younger reactivation of Caledonian structures, are developed in Woodcock (1987b, 1988) and Woodcock *et al.* (1988).

The various interpretations reviewed above have recently been elaborated upon and revised by Smith (1988), Wilkinson (1988) and Wilkinson and Smith (1988), as a result of their comprehensive strain studies in Snowdonia, and by Kokelaar (1988) from an analysis of igneous activity in northern Wales.

Based on Gibbons' (1987) conclusion that the Precambrian rocks of Anglesey and Llŷn represent an amalgamation of terranes, accreted in latest Precambrian to earliest Cambrian times by docking along NE–SW strike-slip zones, it is considered that much of northern Wales is underlain by a similarly heterogeneous basement dissected by steep NE–SW, and probably N–S, shear zones. These shear zones were reactivated, largely as brittle structures, during early Palaeozoic extension and transtension of the basin, and separated fault blocks which could move vertically and laterally independently of one another. In this way, the basement structures strongly governed Cambrian and Ordovician sedimentation patterns and igneous activity. Intermittent differential block movements during basin subsidence, some causing inversion,

may have been responsible for local folding, warping, and tilting, during end-Tremadoc times for example. Subsequently, as a result of the end-Caledonian closure of the Iapetus Ocean (Soper and Hutton, 1984; Soper *et al.*, 1987) the blocks were jostled due to approximately NW–SE compression, again reactivating older faults. Block margin faults which were aligned normal to compression allowed dip-slip displacement, and induced Caledonoid folds and cleavage at and above their tips. On the other hand, faults aligned at an angle with respect to regional compression were reactivated as strike-slip faults, generating transpressive structures in the cover; *en échelon* folds, transecting cleavage and complex patterns of minor faults (Woodcock *et al.*, 1988). This model has therefore combined, modified, and developed the basement control model of Shackleton (1954) and the trans-pressional models of the Welsh end-Caledonian deformation.

ALEXANDRA QUARRY (SH 518561)
R. Scott

Highlights

Alexandra Quarry provides outstanding exposures illustrating Caledonian crustal deformation in the Welsh Cambrian Slate Belt. The Slate Belt of Wales is renowned internationally for its elliptical strain markers, the so-called reduction spots. Small strain markers such as the reduction spots were used in classic early studies of strain measurement. These features can be used to interpret the structure of the Slate Belt, which suffered the highest intensity of Caledonian deformation in North Wales.

Introduction

The overall structure exposed in the quarry is a NE–SW-trending, tight, upright anticline containing the Purple Slate Group in its core, flanked by Dorothea Grit and the overlying Striped Blue Slate Group. This broad structure is complicated by attenuation of the fold limbs and the presence of strike faults.

The geology of the Slate Belt has been investigated from an early stage in the study of Welsh geology because of the wealth of interesting structures exposed in the slate quarries. Sorby (1853, 1856, 1908) interpreted the green spots which characterize some of the slates, as the products of reduction in the sedimentary environment. He used them to estimate shortening perpendicular to cleavage of 75%, combined with 10% volume reduction: this was the first quantitative estimate of strain undertaken. The detailed account of the regional geology is that of Morris and Fearnsides (1926). More recent accounts of regional geology and theoretical structural studies are those of Wood (1969, 1971, 1974), Cattermole and Jones (1970), Tullis and Wood (1975), Wood *et al.* (1976), Roberts (1979), and Wood and Oertel (1980). The Slate Belt has featured prominently in the regional interpretations of Shackleton (1953), Dewey (1969), and Coward and Siddans (1979).

Alexandra Quarry, south-east of Moel Tryfan, was described in detail by Morris and Fearnsides (1926) and lies within the area depicted on the geological map of Cattermole and Jones (1970). The quarry provided a sample locality in the study of Wood and Oertel (1980) in their measurement of strain in the Welsh Slate Belt; and it also appears as a locality in the field guides of Roberts (1979) and Howells *et al.* (1981).

Description

Alexandra Quarry (see Figure 4.3A and B) has been excavated on a variety of levels, the lowest of which are now flooded. Exposure is excellent, although not always accessible. The structure is described at a number of sites within the quarry which are well illustrated by line drawings in Roberts (1979).

Towards the centre of the quarry (around SH 51815613), a large screen of rock separates the north-east end from the lower levels to the south-west (Locality A, Figure 4.3A). When viewed from a position on the lowest part of the track, the south-west face of this rock screen provides a cross-section of the structure perpendicular to the strike. An anticlinal hinge in the Purple Slate is flanked by NW-dipping Dorothea Grit forming the north-west wall of the quarry, and by SE-dipping, tectonically thinned, Dorothea Grit on the south-east side, with Striped Blue Slate above. A prominent greywacke bed showing boudinage occurs on the south-east limb. The screen also contains a vertical dolerite dyke. Like other presumed Ordovician dykes in the Slate Belt it is boudinaged and shows cleavage in its margins.

Although the overall structure displayed in the quarry is anticlinal, strike faults complicate the picture and, unfortunately, the presence of a large

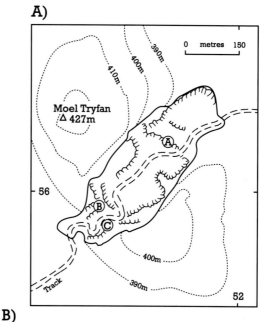

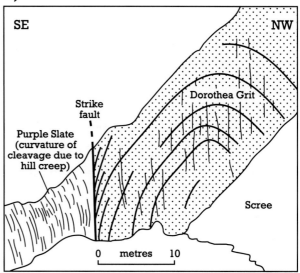

Figure 4.3 Alexandra Quarry. (A) Site map, showing Localities A–C described in the text. (B) Sketch illustrating anticline in Dorothea Grit with a faulted south-east limb. See text for explanation. Locality B of Figure 4.3A.

flooded pit immediately in front of the screen means that several of the interesting structures cannot be directly examined. This can be achieved at the south-west end of the quarry.

On the north-west side of the quarry (at SH 51615599) an anticline can be observed in the Dorothea Grit with the south-east limb faulted out against NW-dipping Purple Slate (Locality B, Figure 4.3A and B). Greywacke beds show quartz-filled tension cracks indicating extension in the outer arc during folding. A few *en échelon* quartz veins and prominent slickensides on bedding surfaces indicate that flexural slip was an important deformation mechanism in the greywackes.

On the south-east side of the quarry (at SH 51655590) south-easterly-dipping Purple Slate forming the face displays a variety of structural features (Locality C, Figure 4.3A). The subvertical, NE–SW striking cleavage forms the main face of the quarry at this location. Grit horizons (<0.05 m thick) in the slate are frequently graded and have load casts. Small-scale folding of these sandstone units is occasionally visible and this mimics the style of the large Slate Belt folds, with their strongly attenuated limbs.

The Purple Slate Group contains green spots of probable diagenetic origin. They have traditionally been termed reduction spots, but the colour of the spots has since been attributed to iron depletion, not reduction (Wood *et al.*, 1976). Sorby (1853) and subsequent workers considered them to be pre-deformation; a conclusion proved by the fact that they are affected by the contact metamorphism of dolerite dykes which have themselves been deformed (Wood, 1973; Wood *et al.*, 1976). During regional deformation, these spots were deformed into ellipsoids whose long (x) axes came to be generally <0.03 m in length. For the Slate Belt as a whole, long axis dimensions in the range 0.01–0.10 m are quoted (Tullis and Wood, 1975).

For initially spherical spots, the long (x) and intermediate (y) direction are contained within the cleavage, and the short axis (z) is perpendicular to the cleavage. This is the case for the majority of isolated spots in Alexandra Quarry, with x-axes plunging steeply to the south-west; indicative of the near-vertical extension recorded throughout the Slate Belt. On joint surfaces which cut the cleavage, the strong flattening of the spots in the plane of cleavage can be observed. Wood and Oertel (1980) recorded ordinary strains (E) (based on 67 measurements) from this quarry, of $(1.00):(0.38):(-0.63)$; that is, the original radii of the sphere have doubled in x, increased by 38% in y and shortened perpendicular to the cleavage by 63%. The x and y dimensions of spots lie oblique to cleavage in cases where the initial shape was irregular. Examples of irregular, and occasional bedding-parallel iron depletion zones can also be observed at this locality.

Interpretation

Alexandra Quarry has been chosen to represent the principal structural features of the Cambrian Slate Belt which suffered the highest intensity of deformation in North Wales. This deformation produced a structural style in sharp contrast to the style exhibited by the volcanics of Snowdonia. Cattermole and Jones (1970) noted the sub-cylindrical nature of large folds in the Slate Belt which differs from the periclinal folds observed in Snowdonia.

Morris and Fearnsides (1926) outlined the main features of the Slate Belt. The NE–SW-oriented belt is separated by boundary faults from the Precambrian and basal Cambrian volcanics to the north-west and the Ordovician slates to the south-east. Internally, the Slate Belt contains numerous fold hinges (dominantly anticlinal) whose attenuated limbs are frequently replaced by strike faults. These essential features are well displayed in Alexandra Quarry. Morris and Fearnsides (1926) described in detail the compressive features of the belt, in which category they placed the folds, cleavage, and two types of strike fault. Their type 1 'slide' type of strike fault is in essence an extreme continuation of the process of flexural slip, whereas the second type is later and post-dates, to some extent, the folding and imposition of cleavage. They estimated horizontal shortening in the belt as >40%, on the basis of evidence provided by a folded bed. They did not, however, recognize the significance of reduction spots, but did indicate that the principal movement of material during deformation was vertically upwards (based on the predominance of anticlinal hinges). Wood (1971, 1974) made a detailed analysis of strain in the Slate Belt; he estimated the tectonic thickness of the succession in the Arfon Anticline to be about double (Wood *in* Rast, 1969).

Cattermole and Jones (1970), although generally following the interpretation of Morris and Fearnsides (1926), described the structural history of the Slate Belt in terms of the deformation phases identified by Roberts (1967). In this scheme, F_1 is the main deformation phase with fold interlimb angles of 65–80° and different fold profiles dependent on lithology. The strike faults were interpreted as high-angle reverse faults.

Morris and Fearnsides (1926) interpreted the development of the Slate Belt using a model involving its compression between the rigid crystalline rocks of Anglesey and the low-lying volcanics of Snowdonia. They suggested that as Snowdonia was progressively driven to the north-west toward Anglesey during the late Silurian, folding, sliding, and cleavage formation were induced in the belt.

Roberts (1979) concluded that the similar folds of the Slate Belt were 'flattened buckle folds which are essentially the result of initial pure shear upon which later simple shearing has occurred'. The interpretation of the deformation was compatible with early suggestions that 'severe flattening was coupled with an essentially upward distention of the sedimentary pile'. The structure of the Slate Belt emphasizes the inhomogeneous nature of deformation in North Wales for which large-scale strain variation, structural position, and lithological control are all likely determining factors.

The present state of research at the site does not allow much advance on the nature of the main deformation other than to confirm the observations of Morris and Fearnsides (1926) and Cattermole and Jones (1970) that the cleavage-parallel movements have been important in the modification of the fold belt. More needs to be known about the relative age and displacement sense of these movements. However, there is no evidence to suggest (cf. Roberts, 1979) that simple shear modification rather than homogeneous flattening of the initial buckle folds has been responsible for the 'similar' folding. The upright cleavage in the context of the other Snowdonia and Anglesey sites, and of other regional research on cleavage in North Wales (Coward and Siddans, 1979; Wilkinson, 1988), does not support the concept of Morris and Fearnsides (1926) that Snowdonia was driven towards Anglesey.

Studies along the length of the Slate Belt have shown the inhomogeneous nature of deformation with tectonic thickening ranging from 50–180%. Maximum extension is coincident with plunge culminations of major folds (and vice versa) (Wood, 1974). Where spots are absent, fabric anisotropy has been used to estimate strain, based on calibration with localities where spots are present (Tullis and Wood, 1972, 1975). The strong agreement between strain, fabric anisotropy and magnetic susceptibility anisotropy has established that slaty cleavage can be accounted for purely by physical rotation (Tullis and Wood, 1972, 1975;

Oertel and Wood, 1974; Wood *et al.*, 1976).

Alexandra Quarry is important as the best and most accessible locality for examining the structural features of the Welsh Slate Belt. The quarry faces provide excellent exposures of characteristic tight anticlinal folds whose limbs are replaced by strike faults. Small-scale strain markers and minor structures enable the intensity and mechanisms of deformation to be established.

The quality of strain data has allowed a detailed understanding of deformation in the Slate Belt: not only of interest in regional terms, but also from a theoretical viewpoint. These perfect triaxial ellipsoids have provided a means to quantitatively evaluate strain during the generation of slaty cleavage since the classic work of Sorby in the nineteenth century, and will remain a data source of international importance. This work has fuelled an international debate (Tullis and Wood, 1972, 1975; Oertel and Wood, 1974; Wood *et al.*, 1976) on the origin of cleavage, of considerable significance in studies of orogenic processes world-wide. The site illustrates high strain-levels typical of the Slate Belt, which suffered the highest Caledonian strains in North Wales. This is an important factor in ongoing studies of the significance of regional strain variations in the Caledonian Orogenic Belt.

Conclusions

The Slate Belt of North Wales is famous for its deformed rocks of Cambrian age. These have been compressed and folded and now take the form of cleaved slates. The area is well known for the studies which have been carried out on these rocks in relation to the extreme compression suffered by Britain during the Caledonian mountain-building phase, around 400 million years ago. Originally-spherical green spots in the muds (so-called reduction spots) are now perfect ellipsoids and have been used in classic studies to quantify the deformation. It has been shown that the sedimentary sequence has been laterally shortened by up to 63% in a north-west to south-east direction, and elongated to become up to double their original thickness. The perfect cleavage planes that characterize these roofing slates are perpendicular to the shortening and are parallel to the upward elongation. This is a classic locality for the study of folds, cleavage and strain related to the Caledonian Orogeny in Wales.

TRUM Y DDYSGL (SH 544518)
R. Scott

Highlights

Trum y Ddysgl provides a superb section illustrating a rare example of overturning of strata on the north-west limb of the Snowdon Syncline. This indicates a high intensity of deformation, and it contrasts markedly with the more open structures and lower strain seen in central Snowdonia to the north-east, along the strike of the fold structures. The site also provides important exposures of a thrust plane, a very rare feature in the Caledonian Orogenic Belt of North Wales.

Introduction

The Trum y Ddysgl site lies on the north-west limb of the Idwal–Snowdon Syncline (Figures 4.1 and 4.4) and provides a contrast in structural style to the Alexandra Quarry, Cwm Idwal and Capel Curig sites. Bedding–cleavage relationships and sedimentary structures in the steeply dipping Cambrian Ffestiniog Beds indicate that the rocks at the base of the cliffs are slightly overturned. The structure is clearly demonstrated (Figure 4.4) by three prominent quartzite horizons. The Cambrian rocks are thrust south-eastwards over Ordovician slate.

Other than the original survey (Ramsay, 1866, 1881), the only detailed description of the site appears in the account of Shackleton (1959, Figure 4.4), who presented a line-drawing illustrating the main structural features. He also presented a measured section through the Ffestiniog Beds. The site lies in an area between Snowdon (Williams, 1927) and the Slate Belt (Morris and Fearnsides, 1926) with their contrasting structural styles. It has been described in the field guide of Roberts (1979).

Description

The site consists of an almost 1 km-long cliff section (Craig Trum y Ddysgl) to the north-east of the summit of Trum y Ddysgl. The overall structure of the locality is best observed from a viewpoint to the north-east on the opposite side of the glacial cirque of which Craig Trum y Ddysgl forms the south-west wall. The site is described by means of a traverse from the north-west to the south-east (Figure 4.4).

The north-west end of the cliffs are formed from the Cambrian Ffestiniog Beds. These are dominated by grey slate with minor, thin (<0.01 m) silt layers defining the bedding. The orientation of bedding is variable, lying close to the core of the Cym-y-Ffynnon Pericline, and tight mesofolds can be seen in places (for example, at 54305208) to which the

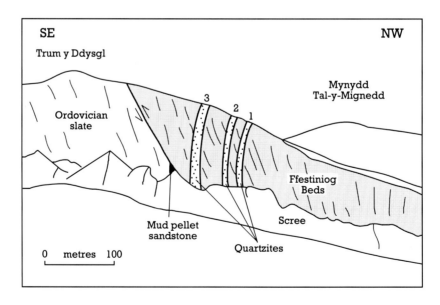

Figure 4.4 View of Trum y Ddysgl looking south-west. Redrawn from Roberts (1979), after Shackleton (1959).

consistently steep NW-dipping cleavage is axial planar. At this locality, cleavage is refracted through more prominent silt layers around the fold hinge and, on the limbs, 0.02 m-thick sand ribs show incipient boudinage. Elsewhere, the tightness of major folds is implied by the small angle between cleavage and bedding.

Traversing along the crags towards the south-east, silt and sand beds become more common until a prominent quartzite bed is reached. This bed (quartzite 1 on Figure 4.4) is the strati-graphically lowest of three prominent quartzite beds up to 20 m thick separated by equivalent thicknesses of shale with thin sandstone beds (Figure 4.4). At the top of the crags, the quartzites dip steeply to the south-east, but at the base they dip steeply to the north-west. Sedimentary structures (cross-lamination and graded bedding) in the shale–silt sequence to the south-east of the quartzite, indicate that the sequence youngs in that direction. The NW-dipping beds at the base of the crags are therefore slightly overturned; here cleavage has the same dip direction but at a lower angle.

Within the thick (sometimes conglomeratic) quartzites, numerous quartz-filled tension gashes can be observed. Some short veins are arranged *en échelon*, while others are more continuous but irregular in orientation; cross-cutting relationships are common. Between the second and third quartzite, 0.5 m-thick sandstone beds in the shale show incipient boudinage, displacement, and quartz veining. In common with the *en échelon* veining in the main quartzite beds, the sense of displacement of the minor boudinaged sandstone beds indicates that the dominant movement of higher levels is toward the south-east.

Further to the south-east along the crags, the Ffestiniog beds are thrust over dark Ordovician slates. The NW-dipping thrust plane 'climbs' the crag in a small gully. The Ordovician slates in the footwall form a monotonous sequence with steeply NW-dipping cleavage, but little indication of bedding. However, at the base of the shale sequence, just below the thrust plane, a small exposed thickness (~2 m) of mud-pellet sandstone, of possible Arenig age (Roberts, 1979), can be seen dipping to the north-west, again indicating over-turning below the thrust.

Interpretation

Whereas the north-east end of the Snowdon Syncline is characterized by open folds of larger wavelength, in the south-west, wavelength becomes less and folds tighten. With this in mind, the site at Trum y Ddysgl has been chosen as a contrast to the structural style displayed in the Cwm Idwal and Capel Curig sites.

In terms of the maximal (major) structures defined by Roberts (1979), the site lies toward the northern termination of the periclinal Cwm Pennant Anticline. Tight folding associated with the Cwm Pennant Lineament is considered to have resulted from relatively intense regional defor-mation associated with the renewal of movement along the line of a synsedimentary fault (Smith, 1988). North of the Cwm Pennant Anticline, the Arfon Anticline becomes the maximal structure adjacent to the Snowdon Syncline. A sharp contrast exists between the strong deformation exhibited by the Slate Belt rocks (for instance, at Alexandra Quarry) on the limb of the Arfon Anticline and the less-deformed volcanic succession of central Snowdonia.

Thrusting is known from only a few rare examples in North Wales; Smith (1987, 1988) has shown that structures interpreted as thrusts by Fearnsides (1910), and Fearnsides and Davies (1944) are, in fact, the product of pre-lithification processes. The variation in fold style between high interlimb angle with upright axial planes and low interlimb angle and more inclined axial planes was attributed to local variations in shear strain by Wilkinson (1987). This model has been developed by Wilkinson (1988) and Smith (1988) who attribute the lower angle of axial planes of some folds to propagation above the tip lines of thrusts which may flatten into cover detachments, the cover–basement interface, or faults within the basement. The Trum y Ddysgl site represents a very rare location where one of these thrusts extends up to exposed levels.

The site provides an excellent example of the fold style and the more intense deformation which characterize the south-west part of the Snowdon Syncline, more particularly on its north-west limb. The section displays overturning, seen in bedding–cleavage relationships and proved by sedimentary structures. Cambrian strata are thrust south-eastwards over Ordovician slates on a NW-dipping thrust plane. Kinematic indicators in quartzite beds of the hanging wall are consistent with the south-easterly overthrusting indicated by strati-graphical relationships.

The regional variations in structural style are central to ongoing reinterpretation of the Caledonian Orogenic Belt, both in North Wales and in Britain as a whole. Particular significance is now attached to the relationship between strain intensity, reflected in fold style, and basement faulting, and this site is of special importance because of the coincidence of high strain and thrust faulting at a single locality. The full significance of these features is not yet certain and this site is likely to attract considerable further study and interpretation.

Conclusions

At Trum y Ddysgl, folding and thrusting by low-angle faults, which were formed during the Caledonian mountain-building episode, affect rocks of the Cambrian and Ordovician periods. Cambrian strata are overturned and are thrust south-eastwards over the younger Ordovician rocks, although such thrusts were a rarity in the Caledonian terrane of Wales. The intense deformation seen here, contrasts with other areas of Snowdonia where the effects of the orogeny were less pronounced.

CWM IDWAL (SH 638588–649607)
W. R. Fitches

Highlights

The Idwal Syncline is the best known and most studied major Caledonian fold in Wales. Cwm Idwal is excavated along the axis of the syncline; it provides outstanding exposures of the fold in plan and profile.

Introduction

Cwm Idwal (Figure 4.5), a glacial cirque, is cut along the hinge zone of the Idwal Syncline, which is a major end-Caledonian fold. The

Figure 4.5 Cwm Idwal, Gwynedd. The right- and left-sloping slabs above the central scree form the syncline hinge of one of the major Caledonian fold structures in Snowdonia, in Ordovician sediments and volcanics. View to south-west, cliff is approximately 300 m high. (Photo: S. Campbell.)

rocks deformed by the structure are products of Ordovician (Caradoc Series) volcanism and sedimentation and belong to the upper part of the Cwm Eigiau Formation, the Lower Rhyolitic Tuff Formation, and the lower part of the Bedded Pyroclastic Formation (Figure 4.6; Howells *et al.*, 1981). The various rocks and processes which produced them are outlined in field guides to Cwm Idwal (Roberts, 1979; Howells *et al.*, 1981) and the site is encompassed by the BGS 1:25 000 Geological Special Sheet for Central Snowdonia.

Description

The Idwal Syncline is a component of the Snowdonia Synclinorium – see Roberts, 1979; Figures 4.1 and 4.2. Its profile is exposed in the back wall of the cirque, where it is seen to be an open to close structure with a NE–SW (Caledonoid) axial plane inclined very steeply to the north-west, while the hinge line is nearly horizontal. Bedding planes in the limbs dip at about 50° toward the trough of the fold. The axial trace continues to the north-east to Llyn Idwal, then turns N–S to pass through the Idwal Cottage Youth Hostel area before ascending the slopes of the Carneddau. In the Idwal Cottage area, the fold tightens and becomes more asymmetrical as the western limb steepens more than the eastern limb, and the southerly plunge becomes moderate to steep. The tightening and increase in plunge intensify toward the closure of the fold in the Carneddau, probably because there the layered rocks were compressed against the Pen-yr-Ole-Wen rhyolite plug (on the north side of Nant Francon) (Wilkinson, 1988, p. 80).

The site exhibits a variety of small-scale structures, several of which have been recently investigated by Wilkinson (1988) during his analysis of strains in Snowdonia. In particular, he used the siliceous concretions in the Pitts Head Tuff and the accretionary lapilli in the Lower Rhyolitic Tuff to measure the amounts of shortening and extension which took place during deformation of the rocks.

The site encompasses an area of several square kilometres so, for descriptive convenience, six separate localities (A–F), containing structures representative of those found in various parts of the site, are described in the following sections (Figure 4.6).

Locality A: South-east limb of Idwal Syncline: Idwal Slabs [SH 645589]

The rocks comprising the Idwal Slabs and adjacent crags are massive-bedded, acid ash-flow tuffs, volcanic breccias, tuffs, and interbedded sandstones and mudstones of the lower part of the Lower Rhyolitic Tuff (Howells *et al.*, 1981, p. 61). Layering dips to the north-west at moderate angles in the south-east limb of the Idwal Syncline, and this has been exploited by erosion to form the surfaces of the Slabs. Cleavage is poorly developed in the massive, rigid rocks, but can be discerned as a result of the weathering out of phyllosilicates and fine volcanic fragments, which are weakly aligned following the tectonic fabric. In places, some of the larger volcanic fragments, up to 0.15 m across, have a preferred alignment in cleavage, although the majority of clasts are aligned in bedding.

The rocks are cut by numerous veins of quartz, accompanied locally by chlorite, which range in thickness from less than one millimetre to about 0.25 m. Several veins are parallel, or nearly parallel, to layering and some of these comprise quartz fibres elongated normal to vein walls. These bedding-parallel veins are commonly slickensided, the striations plunging down the dip of the vein. A second type of quartz vein, accompanying the other type, or occurring independently, is horizontal or inclined gently to the south-east. Several large examples of this second type crop out high up on the Idwal Slabs. Quartz fibres in these flat-lying veins are also elongated normal to the vein walls and are parallel with the striations on the bedding-parallel veins. The flat-lying veins are seen locally to cut those of the other type. Rarely, the flat-lying veins are gently folded in the cleavage.

Locality B: Crags immediately south-west of Idwal Slabs [SH 644589]

The crags adjacent to the footpath and above the Idwal Slabs are formed of beds of upper Lower Rhyolitic Tuff Formation, comprising well-bedded sandstones, siltstones, tuffs, and tuffites (Howells *et al.*, 1981, p. 61). Here, cleavage is more strongly developed than in the Idwal Slabs, because of the better alignment of feldspars and volcanic fragments in the tectonic fabric. Large volcanic clasts and brown-weathering carbonate concretions, between 0.10 m and 1.5 m across, mostly remain aligned parallel with bedding, but some smaller

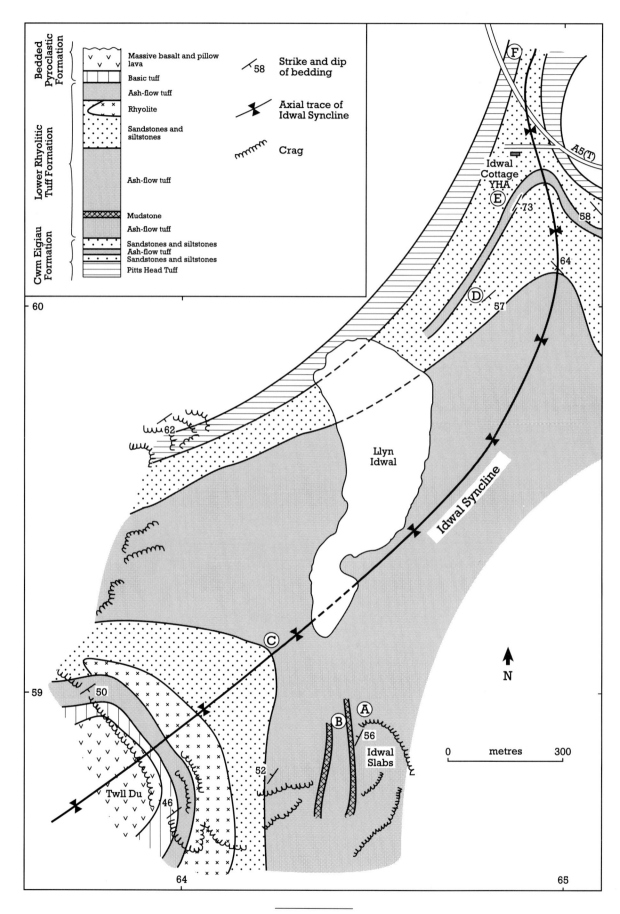

Bedded Pyroclastic Formation

Massive basalt and pillow lava

Basic tuff

Lower Rhyolitic Tuff Formation

Ash-flow tuff

Rhyolite

Sandstones and siltstones

Ash-flow tuff

Mudstone

Ash-flow tuff

Cwm Eigiau Formation

Sandstones and siltstones
Ash-flow tuff
Sandstones and siltstones
Pitts Head Tuff

Strike and dip of bedding

Axial trace of Idwal Syncline

Crag

Idwal Cottage YHA

A5(T)

Llyn Idwal

Idwal Syncline

N

Idwal Slabs

Twll Du

0   metres   300

ones have been apparently rotated towards the cleavage.

Locality C: Syncline core [SH 642591]

The well-bedded rocks at (B) dip at progressively shallower angles to the north-west into the hinge of the Idwal Syncline, forming a line of low crags which descend from locality B towards the southern end of Llyn Idwal. Clean, fresh exposures of the rocks are found in the beds of streams descending to the lake and cutting through the crags.

Cleavage is well-developed in these rocks of the fold hinge, and is formed by the strong preferred alignment of quartz grains, quartz pressure-fringes on feldspars, and phyllosilicates. The rocks also contain scattered, dark-grey accretionary lapilli, pea-sized clots of volcanic ash which were originally nearly spherical, but are now deformed into triaxial ellipsoids, flattened parallel to the cleavage and extended vertically. Volcanic clasts and carbonate concretions also commonly show strong alignment in the cleavage; contrasting with most other localities where the majority remain elongated parallel to the bedding.

Locality D: North-west limb of syncline [around SH 646601]

The Pitts Head Tuff and the sandstones, siltstones, and acid tuffs above, belonging to the Cwm Eigiau Formation, are well exposed in the north-west limb of the Idwal Syncline, in a broad strike-parallel ridge which extends from the northern end of Llyn Idwal toward the Idwal Cottage Youth Hostel. Bedding dips steeply to the south-east at angles of up to 80°, and cleavage dips at about 70°NW. Cross-bedding in the sandstones shows that the beds are the right way-up and that the structure is upward-facing.

Siliceous concretions (lithophysae) are common in the upper part of the Pitts Head Tuff. These are pea- to tennis ball-sized, white-weathering masses of silica, or very fine quartz, which were precipitated from solutions migrating through the volcanic pile soon after its accumulation. The originally nearly

spherical concretions have been slightly flattened in cleavage and extended vertically during deformation.

In places along the ridge, the Pitts Head Tuff shows well-developed columnar jointing formed during cooling and contraction of the volcanic pile. The columns were originally set perpendicular to layering, but have been realigned by compression in the Idwal Syncline and now lie at angles of about 60° to layering and plunge at moderate angles to the north-west. These deformed columns afford the opportunity for the measurement of strain in the Idwal Syncline. As at the Idwal Slabs (above), this ridge contains numerous examples of quartz veins. Some are parallel with bedding, while others with a flat-lying attitude, form ladder arrays nearly parallel with bedding.

Locality E: Honestone Quarry [SH 648602]

The long, ravine-like quarry immediately above the Youth Hostel cuts along the strike of inter-bedded, light-coloured, fine-grained tuffs and dark mudstones of the Eigiau Formation. Bedding dips very steeply to the south-east, in the north-west limb of the Idwal Syncline, and according to cross-bedding in some tuffs, youngs to the south-east. Cleavage is nearly vertical and is unusually strong due to the high degree of phyllosilicate alignment, which is sufficiently intense to give some of the deformed mudstones a phyllitic fabric, characterized by shiny surfaces.

Locality F: A.5 road-cutting and adjacent crags [SH 649606]

A series of crags some 15 m north-east of the road exposes well-bedded volcanogenic sandstones which lie immediately above the Pitts Head Tuff. The sandstones, dipping about 70°SE, contain 0.1–0.5 m-thick layers, crowded with 10–40 mm brachiopod shells preserved in brown-weathering carbonate. These shells have been strongly deformed in the steeply dipping cleavage. On steep surfaces striking at right-angles to the cleavage, the originally gently curved shells are seen to be almost isoclinally folded.

In the cutting on the north-east side of the road, the upper parts of the Pitts Head Tuff show the very strong planar alignment of *fiammé* (pumice fragments) and other volcanic ejecta caused by compaction and welding during accumulation of the hot ash pile; this fabric is mostly parallel with

Figure 4.6 Cwm Idwal. Geological map showing the positions of Localities A–F described in the text.

bedding in the overlying sedimentary rocks. Unlike those sedimentary rocks, the welded tuffs are often uncleaved, due to their greater rigidity during deformation. However, cleavage is developed in places, usually in narrow, 0.5 m-wide zones. Locally in the tuffs, particularly rigid layers have been boudinaged, individual boudins becoming separated by 0.1–0.2 m-long quartz lenses.

In the cutting on the south-west side of the road, there are several quartz veins, mostly less than 1 m in length, which dip moderately to steeply to the south-east. These veins are arranged *en échelon* in two tension gash arrays, the larger of which is over 10 m in length. The sense of shear indicated by these arrays is top side to the south-east.

The Idwal Syncline as exposed in Cwm Idwal is a major fold with a steep NE–SW axial plane and nearly horizontal plunge. The south-east limb is represented at the Idwal Slabs (Locality A), the north-west limb at Locality D, and the core at Locality C. Toward the north-east the fold tightens, the axial plane swings to a more nearly N–S strike and the plunge becomes steep to the south, probably because the rocks were deformed against the rigid Pen-yr-Ole-Wen Rhyolite (Wilkinson, 1988). The fold is typical of those produced by the main deformation phase. It offers particularly fine, three-dimensional exposures to study the full geometry of such a major fold.

The exposures, the variety of lithologies and the presence of deformed strain markers provide an unrivalled opportunity to examine the relationships of cleavage to a main-phase fold. Cleavage is more or less axial planar to the syncline in Cwm Idwal, striking NE–SW and dipping very steeply to the north-west. Wilkinson (1988) reports that it transects the northern end of the structure in a clockwise sense in the flanks of the Carneddau, north of the site.

The character of the cleavage within the site is variable, probably because of the variations in lithology. Cleavage is commonly absent or barely discernible in the strongly welded parts of the Pitts Head Tuff (for example, Locality F). The ash-flow tuffs, breccias and coarse volcanogenic sandstones usually take a feeble cleavage; a spaced type with little grain alignment. In parts of the volcanic rocks, however, the cleavage becomes penetrative, as the volcanic clasts, quartz pressure fringes on feldspars, and new phyllosilicates take on a strong preferred alignment. The strongest cleavage within the site is found in the Honestone Quarry (for example, Locality E) where phyllosilicates in the original mudstones and fine-grained volcanic and volcaniclastic rocks are aligned sufficiently uniformly to give an almost phyllitic cleavage.

There are various strain markers within the site: originally nearly spherical accretionary lapilli (for example, Locality C) and siliceous concretions (for example, Locality D), brachiopod shells (for example, Locality F), and columnar joints which were originally elongated normal to layering in the Pitts Head Tuff (for example, Locality D). Wilkinson (1988) has used the accretionary lapilli and concretions to show that strains in Cwm Idwal are nearly plane strain: the flattening in cleavage is almost compensated for by vertical extension while there was little change in dimensions horizontally in the cleavage.

There are a number of small-scale structures exposed at the site, related to the development of the main-phase fold. These have not been studied previously but offer considerable research potential. Veins are common but appear to have attracted no comment in published literature. Of particular interest are the veins aligned parallel with bedding and those which are flat lying (for example, Localities A and D). These two types of vein are considered here to be products of flexural slip which accompanied the development of the Idwal Syncline. The bedding-parallel veins represent slip surfaces along bedding, while the others are more or less contemporaneous tension gash arrays. The slip-senses, deduced from rare *en bayonet* bedding-parallel veins, the *en échelon* arrangement of the other veins (Figure 4.7B), and the down-dip striations on bedding-parallel veins, are consistent with displacement resulting from accommodation during folding (Figure 4.7B).

Interpretation

The folding of some tension-gash veinlets (Locality A) is interpreted as the effect of flexural slip during initial buckling, as illustrated in Figure 4.7B. The observation that these veinlets are themselves folded, indicates that cleavage development outlasted parts of the flexural slip history of the fold. The boudins (Locality F) are interpreted as examples of inverse boudins (cf. Ramsay, 1983, Figure 3B); quartz veins which had been formed at the necks of early, square-ended boudins acted as more rigid layers during later, more ductile stages of extension, so that the sites of necking moved

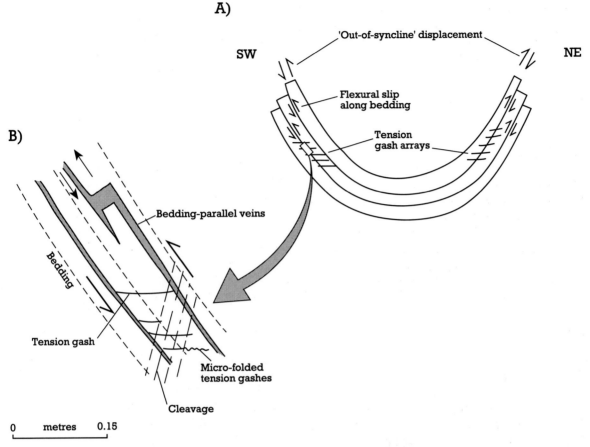

Figure 4.7   Cwm Idwal. (A) 'Out-of-syncline' flexural slip and tension gash arrays in the Idwal Syncline. (B) Combination of *en bayonet* bedding-parallel veins and tension gashes, south-east limit of Idwal Syncline (Locality A).

into the boudins which had been formed early in the progressive deformation.

The Idwal Syncline is one of the best-known Caledonian structures in Britain thanks to the superb exposures provided in the floor and walls of Cwm Idwal. The scale of the features, and the high level of exposure here, present an outstanding opportunity for detailed structural studies including three-dimensional strain variations within a major Caledonian fold, developed in the main phase during the late Silurian–early Devonian. The fold has the NE–SW trend characteristic both of Snowdonia and of the British Caledonian fold belt. The Idwal Syncline characterizes the gentler style of deformation and lower levels of strain typical of central Snowdonia, contrasting with the high strains of the Slate Belt. There is marked variation in cleavage development within the site from very weak in parts of the Pitts Head Tuff to strong and phyllitic in the mudstones of the Honestone Quarry. Other minor structures such as volcanic

lapilli and deformed fossils are valuable as strain markers and have been used in recent regional studies. The locality provides the best exposed structure of its kind in central Snowdonia, with great potential for future studies.

Conclusions

The scale and three-dimensional nature of the rock exposures at Cwm Idwal provide unrivalled opportunities for detailed studies of a major fold formed during the Caledonian mountain-building period. Small-scale structures associated with the fold, including cleavage (fine, closely-spaced, parallel fractures), deformed fossils, concretions, and mineral veins are exposed at many localities, and these enable detailed studies of the nature and intensity of strain and its variation throughout the folds to be carried out. This site affords opportunities to study, in three dimensions, a major structure that may be assigned to the main phase of Caledonian deformation around 400 million years before the present during the late Silurian or early Devonian.

CAPEL CURIG (SH 707563)
R. Scott

Highlights

The crags east of Dyffryd Mymbyr contain excellent examples of deformed accretionary lapilli which provide an important measure of the amount of crustal strain which affected the volcanic rocks of Snowdonia during the Caledonian Orogeny. The deformed lapilli also give an indication of the nature of Caledonian strain in this area; the site is therefore important as part of a network representing the regional structural pattern.

Introduction

The site shows outcrops of the Dyffryn Mymbyr Tuff; the highest tuff unit of the Capel Curig Volcanic Formation (Caradoc Series). This member only occurs on the north-west limb of the periclinal Capel Curig Anticline (Howells *et al.*, 1978) and is characterized by beds rich in whole and fragmented accretionary lapilli indicative of subaerial deposition. The deformed lapilli record the state of strain in this component fold of the Snowdonia Syncline – see Figures 4.1 and 4.2. The lapilli tuff at Capel Curig has become a standard for the practical illustration of strain measurement in undergraduate teaching and in numerous textbooks (for example, Ramsay, 1967; Ramsay and Huber, 1983).

The Capel Curig district was first investigated in a systematic fashion by Jukes, Aveline, and Selwyn who undertook the primary survey, started in 1848, with maps and sections published by the Geological Survey (1851–55). The structure and stratigraphy of the district were also outlined in the North Wales memoir (Ramsay, 1866, 1881). The first detailed map and description was that of Williams (1922). In recent years, the district around Capel Curig has been incorporated in regional strain studies (Siddans, 1971; Coward and Siddans, 1979; Wilkinson, 1987, 1988). A number of studies have concentrated on volcanological

and stratigraphical aspects of the Capel Curig Volcanic Formation (Francis and Howells, 1973; Howells *et al.*, 1978, 1979) and descriptions of the area have appeared in the field guides of Roberts (1979) and Howells *et al.* (1981).

Description

The site consists of a single exposure of the Dyffryn Mymbyr Tuff, located 200–300 m to the ENE of Dyffryd Mymbyr. The exposure forms a prominent crag, 10–20 m high, overlooking an area of large fallen blocks.

The bedded Dyffryn Mymbyr Tuff lies on the north-west limb of the NE–SW-trending periclinal Capel Curig Anticline. The SW-plunging closure of this fold is well displayed in the sandstones underlying the Capel Curig Volcanic Formation on Creigiau'r Garth to the south. At the GCR site, the tuffs dip at approximately 20° to the north-west, the bedding being defined by colour variations and by different concentrations of the ellipsoidal accretionary lapilli. On a small scale, minor irregularities can be seen in bedding orientation. The higher parts of the crag are composed of paler crystal tuff with few lapilli.

Cleavage is somewhat variable in intensity and irregular in orientation, refracting between lithologies. The general dip is 70° to the north-west, and the steep face of the crags runs approximately parallel to the 050°–060° strike. Large cleavage surfaces >0.15 m apart define units containing weaker cleavage surfaces which are discontinuous and occasionally anastomose around the lapilli.

The best examples of the strained accretionary lapilli can be observed at the base of the crags on either side of a small wall. Individual lapilli are near-perfect ellipsoids. They have x dimensions from 1 to 25 mm which on cleavage (xy) planes, pitch 70° from the north-east, illustrating the steep nature of extension. Some, although by no means all, of the larger crystal fragments in the surrounding tuff are also elongated parallel to the long axes of the lapilli. The y axis of the lapilli is parallel to the gently plunging axis of the south-west end of the Capel Curig Anticline. Large joint surfaces intersecting the face of the crags produce planes approximating to the (xz) axes in which flattening in the plane of cleavage can be observed. Overall, axial ratios in the lapilli ($x>y>z$) approximate to 4:3:1.

Interpretation

The Dyffryd Mymbyr site has been chosen as a representative location displaying excellent strain markers and showing the intensity of deformation in Snowdonia.

The Capel Curig Anticline is a component fold of the Snowdonia Syncline and it lies within the central zone of that structure. It is one of the best examples of periclinal folding in Snowdonia. The fold typifies the north-east end of the synclinorium, being an open, NE–SW-trending symmetrical structure. Towards the south-west, folds decrease in wavelength and interlimb angle – see Trum y Ddysgl. All folds are characterized by the absence of meso-scale folding and an absence of hinge-zone thickening.

The arcuate nature of the Snowdonia Syncline (convex to the north-west) was initially interpreted as a primary feature of the deformation by Shackleton (1953) and, later, by Dewey (1969). In contrast, Helm *et al.* (1963) suggested that re-folding was responsible. This was disputed by Coward and Siddans (1979) who argued for (NW–SE) compression against the rigid indentor of the Berwyn Hills. Campbell *et al.* (1985) favoured this latter model, but suggested that the indentor was the NW-dipping concealed extension of the Tan y Grisiau microgranite. Strain has been measured in the volcanic rocks of Snowdonia (Siddans, 1971; Roberts and Siddans, 1971; Coward and Siddans, 1979; Wilkinson, 1987, 1988) using a variety of volcanogenic markers (siliceous nodules, rhyolite clasts, tuff clasts, and accretionary lapilli). The compilation of strain data by Coward and Siddans (1979) is incompatible with refolding and, therefore, disagrees with the interpretation of Helm *et al.* (1963). Wilkinson (1987) showed strain in Snowdonia to be heterogeneous, but approximating overall to plane strain with a vertical extension rarely exceeding 130% in tuffs.

The Dyffryd Mymbyr site displays excellent examples of strained accretionary lapilli. The steep, north-westerly plunge of the *x* axis, sub-horizontal NE–SW *y* axis, and shallow south-easterly plunging *z* axis shown by these lapilli are typical throughout Snowdonia. In common with the Lower Palaeozoic succession throughout North Wales, this indicates vertical extension in response to NW–SE compression during the main phase deformation, although the extent of deformation varies with location and lithology (cf. Alexandra Quarry). The lapilli may indicate a higher state of strain and a more pronounced flattening deformation in the volcanics than that shown by other strain markers because of the likelihood of a higher volume loss in these other lithologies than in the more massive volcanics (Wilkinson, 1987).

The regional variations in strain values and structural style have still to be incorporated in a widely accepted tectonic model. The variety of models: basement control (Shackleton, 1953; Dewey, 1969); 'thin-skinned' tectonics (Coward and Siddans, 1979; Campbell *et al.*, 1985); oblique-slip (Woodcock, 1984b)) were summarized by Wilkinson (1987), whose own work emphasizes the heterogeneous nature of strain in the Ordovician volcanic sequence. This heterogeneity has recently been considered in the models of Wilkinson (1988), Smith (1988) and Wilkinson and Smith (1988). They suggest that the style of structures and the intensity of strain developed in the Palaeozoic cover was determined by the orientation and distribution of basement faults which were active during sedimentation and deformation.

Conclusions

The tuffs at this locality contain accretionary lapilli (originally, spherical hailstone-like accumulations of volcanic ash). These now perfect ellipsoidal objects are excellent indicators with which the Caledonian strain within the Ordovician volcanic rocks of Snowdonia can be measured. Their degree of distortion makes it possible to assess the actual amount of tectonic deformation (crustal shortening) to which Snowdonia was subjected during the Caledonian mountain-building episode, around 400 million years before the present.

Strain measurement has played an important role in interpreting the structure of the Caledonian Orogenic Belt of North Wales, and there is ongoing research in this field. This site lies within the Capel Curig Anticline, probably the best example of the periclinal folds of Snowdonia, and it illustrates the open structural style of the north-eastern end of the Snowdonia Synclinorium, contrasting markedly with the site at Trum y Ddysgl where deformation was more acute.

TAN Y GRISIAU (SH 683454)
R. Scott

Highlights

This locality displays excellent examples of deformed contact-metamorphic spots in cleaved Tremadoc siltstones and pelites. The contact spots provide a means of dating the Tan y Grisiau microgranite intrusion in relation to the deformation of North Wales that was imposed during the Caledonian Orogeny. In addition, the spots and cleavage are important in the measurement and analysis of the strain imposed in southern Snowdonia during the orogeny, and this has an important bearing on the postulated 'Tremadoc Thrust Zone'.

Introduction

The site provides an example of deformed metamorphic spots in Tremadoc Series siltstones and pelites within the contact aureole of the Tan y Grisiau microgranite. The microgranite, which lies 700 m south-east of the site, has a roof area, estimated by Campbell *et al.* (1985) to measure about 10 km by 5 km, dipping beneath the site at about 20°. The aureole extends at least 1 km on this north-west side. Dark metamorphic spots are flattened in the plane of the north-dipping cleavage and are extended down dip.

Areas adjacent to the Tan y Grisiau intrusion have been the subject of numerous structural studies, of which the work of Fearnsides (1910) and Fearnsides and Davies (1944) are notable early examples. To the south-west of the microgranite lies the Tremadoc Thrust Zone, a band of crush belts and high strain which may be olistostromic. A detailed petrological study of the granite was presented by Bromley (1963) who also contributed to a series of later papers (Bromley, 1969, 1971; Lynas, 1970a, 1973; Fitch *et al.*, 1969), which were concerned with the structural relationships of the granite and its host rocks. The interpretation of the flat-lying cleavage favoured by Lynas (1970a) was combined with the thrust model of Fearnsides and Davies (1944) in the general structural interpretation of Coward and Siddans (1979).

Recent work by Campbell *et al.* (1985) has modified the general model of Coward and Siddans (1979), and has confirmed the view that low-angle cleavage around Tan y Grisiau is equivalent to the upright main cleavage elsewhere in North Wales. Smith (1987, 1988) has cast doubt on previous interpretations of low-angle discordances to the west of Tan y Grisiau. Instead, he suggests that the Tremadoc Thrust Zone (Fearnsides, 1910) may be an olistostrome.

Description

The Tan y Grisiau GCR site consists of an exposure of metasediment illustrating representative examples of the deformed contact spots. Within the exposure, Tremadoc sediments dip to the north at ~45°, with a single cleavage dipping at a slightly steeper angle in the same direction. Numerous, black contact-metamorphic spots appear throughout the sediments and different concentrations of spots help define the bedding. The spots are mainly oval in shape and generally have a maximum diameter <0.01 m, although some approach 0.02 m. Cordierite was the original mineral forming these rounded spots which now have a retrogressive mineralogy of chlorite and sericite. Occasional angular spots which are lath- or diamond-shaped may have had andalusite as a precursor.

Strain can be estimated using a combination of cleavage and joint surfaces on which various sections of the strain ellipse can be measured. The spots here are flattened in the cleavage and have x-axes which plunge down the cleavage surface toward the north. A grain-shape fabric in the matrix has the same orientation. Joint surfaces in a variety of orientations allow an accurate picture of the strain ellipsoid to be obtained. An average axial ratio ($x:y:z:$) of 1.72:1:0.67 has been calculated at the site using 30 spots (Smith, 1988).

Several thin (<0.01 m) veins cross-cut the sediments and some possess symmetrically disposed colour zoning, produced by hydrothermal alteration, which may extend up to 0.04 m into the surrounding rock. Occasionally these veins contain euhedral quartz, calcite, and minor pyrite. The veins, and the retrogressive mineralogy of the spots, are the consequence of an expulsion of volatiles from the granite during the latter stages of its crystallization.

Interpretation

Two aspects of the geology of the Tan y Grisiau area have provoked controversy in the literature:

1. the relative age of the microgranite intrusion with respect to the regional deformation of the surrounding rocks; and
2. the age of the low-angle cleavage in the aureole and elsewhere on the southern margin of Snowdonia.

Both these age relationships are crucial to the interpretation of the structural development of the area between Snowdonia and the Harlech Dome. The chosen site provides an example of the orientation relationships and strain data available for such investigations.

The accepted age of the Tan y Grisiau microgranite, with respect to deformation, has progressively changed during the course of a prolonged period of investigation. Early workers considered it to be post-tectonic (Jennings and Williams, 1891), whereas Fearnsides and Davies (1944) considered it to be post-cleavage, but to pre-date the Tremadoc Thrust (Fearnsides, 1910). Shackleton (1953) considered the intrusion to be truly synorogenic. However, more recent research has demonstrated that the intrusion pre-dated the cleavage because the contact-metamorphic spots are deformed within the plane of the main cleavage (Bromley, 1969; Coward and Siddans, 1979), an interpretation confirmed by a minimum age of 477 ± 20 Ma obtained for the granite by Fitch *et al.* (1969).

The aureole of the microgranite is characterized by a low-angle, northerly dipping cleavage which is developed only in Cambrian and Ordovician strata along the northern flank of the Harlech Dome. Lynas (1970a, 1973) interpreted this flat-lying cleavage as the product of a deformation which preceded that forming the main cleavage elsewhere. Coward and Siddans (1979) suggested that this fabric was related to the development of the Tremadoc Thrust Zone. However, the interpretation of this flat-lying cleavage as a low-angle manifestation of the steeply dipping main phase (that is, late Silurian–early Devonian) cleavage elsewhere (Bromley, 1971), has been confirmed by recent work (Campbell *et al.*, 1985; Smith, 1987, 1988).

Campbell *et al.* (1985) reassessed the model of Coward and Siddans (1979) and, while they still preferred a 'thin-skinned' interpretation of structural evolution, modified it so that the subsurface extension of the Tan y Grisiau microgranite played a dominant role in thrusting. In their model, the microgranite body acted as a rigid block over the roof of which the bulk of Snowdonia was ramped during the main Caledonian deformation. This provided an explanation for both the shallow dip of the main cleavage and the northerly dipping extension direction indicated by the contact spots and mineral grain elongation. In addition, the low angle between cleavage and bedding was thought to have facilitated dislocation along the Tremadoc Thrust Zone where shear-strain was at a maximum.

Recent work by Smith (1987, 1988) has reassessed the evidence for the existence of the Tremadoc Thrust Zone and favours a pre-deformation explanation for the features previously attributed to thrusting; features such as crushing, faulting, bed repetition, and high strain associated with the zone. A strain study, which included investigation of the deformed contact spots of the microgranite aureole, indicated relatively low strain with the exception of a narrow zone of intense prolate strains in the Rhyd area, a 7 km-long strike to the south-west. Smith (1987, 1988) considered these unusually high strains to be related to compression against the rigid subsurface extension of the microgranite, but high strains being achieved without detachment along a specific thrust plane; an assessment recently confirmed by radiometric methods.

In addition, the siltstones and shales have a 'low-angle' cleavage found extensively along the northern flank of the Harlech Dome. This cleavage and the related Tremadoc Thrust Zone have been the subject of some controversy. It is now agreed that the cleavage at the site is a variation on the main-phase regional cleavage whose low angle of dip is a local deflection related to the presence of the underlying microgranite intrusion. One interpretation (Campbell *et al.*, 1985) sees the low angle and local high strain as results of a thrust ramp which transported Snowdonia south-eastwards over the rigid block during the main deformation phase. However, Smith (1987) denies the presence of any discrete thrusting, but accepts that the angle of cleavage and its intensity has been controlled by the presence of the microgranite. Although the measurement of the deformed contact spots at this site have provided further important data for the nature of the Caledonian strain, measurements of similar spots in the contact aureole over a wider area would enable the effect of the pre-deformation intrusion to be seen in a wider context.

Conclusions

This locality shows excellent exposures of Tremadoc (early Ordovician Period) siltstones and shales which have been affected by baking by a later igneous intrusion, the Tan y Grisiau microgranite. Contact spots, a product of the baking, have developed within the altered zone (aureole) around the Tan y Grisiau microgranite intrusion. These spots may be seen to be deformed, which is evidence of tectonic deformation after the emplacement of the microgranite. The deformed spots have played an important role in resolving the debate about the age of the intrusion relative to the formation of the main Caledonian cleavage.

The microgranite was emplaced around 470 million years before the present. This is consistent with a date of around 400 million years for the main Caledonian mountain-building event, including the low-angle cleavage, which deforms the spots in the granite aureole. The unusual cleavage here has been interpreted as due to Snowdonia being pushed (thrust) southwards over the buried microgranite mass.

OGOF GYNFOR
(SH 37779476–37939500)
D. E. B. Bates

Highlights

At Ogof Gynfor the Precambrian Mona Complex of Anglesey is overlain unconformably by Ordovician conglomerates and cherty shales. The dramatic folding and faulting of the two sequences is of great importance in the controversy over the stratigraphical, structural, and metamorphic relationships between the two units, which represent one type of relationship between basement and cover in the Welsh Caledonides.

Introduction

The older rocks at Ogof Gynfor (Figure 4.8) consist of siliceous Gwna Mélange of the Precambrian Mona Complex. It contains a large mass of quartzite which may form a particularly large block within the mélange. The Ordovician conglomerates, of the Torllwyn Formation of the Arenig Series, rest unconformably on this basement and are succeeded disconformably by the

Caradoc Gynfor Shales. The Ordovician sequence has been strongly folded, into a series of four synclines and three anticlines, with dips up to the vertical. Both rock units have then been cut by reverse faults, giving some of the folds the geometry of hanging-wall anticlines and footwall synclines; lower-angle thrust splays are present, and finally there are steep northerly-dipping normal faults. A schistosity pervades the Mona Complex and a slaty, or spaced, cleavage the Ordovician rocks.

Matley (1899, p. 648) first described the section in detail, and used it to demonstrate the existence of a sub-Ordovician unconformity in Anglesey, and the presence of thrusting. It was also described and figured by Greenly (1919). Shackleton (1954) drew attention to the basement to cover relationships shown by the faulting, and Bates (1968) recognized the Arenig–Caradoc disconformity, and gave further description (Bates, 1972, 1974).

More recently, controversy has centred on the relationship between the Mona Complex and the Ordovician. Barber and Max (1979) have claimed that, contrary to earlier workers, both sequences were affected by a single deformation event, placing them in their Cemlyn Tectonic Unit.

Description

The sequence from Llanbadrig Point to the south side of Ogof Gynfor is formed of Gwna Mélange. At the south side of the inlet (SH 37869475) a fault with a steep northerly dip downfaults the Ordovician Torllwyn Formation to the north. It rests here unconformably on the Gwna Mélange, but the surface of the unconformity is inaccessible in the cliff. Above is a small quarry in the Arenig conglomerates, with poorly preserved brachiopods. Within the inlet are several fault-bounded masses of Gwna Mélange, Arenig Torllwyn Formation and Caradoc Gynfor Shales (Figure 4.8). On the north side of the inlet a major, vertical, WNW–ESE-striking fault separates this complex from a high ridge of Gwna Mélange and quartzite. This ridge is terminated to the north by another vertical fault, which downthrows the succession once more to bring the Ordovician conglomerates to sea-level.

The cliffs from here (SH 37829484) to the north end of the section expose the irregular unconformity between the Gwna Mélange and the overlying grits. The mélange contains a marked penetrative cleavage, which has a similar steep attitude to the spaced cleavage in the grits above.

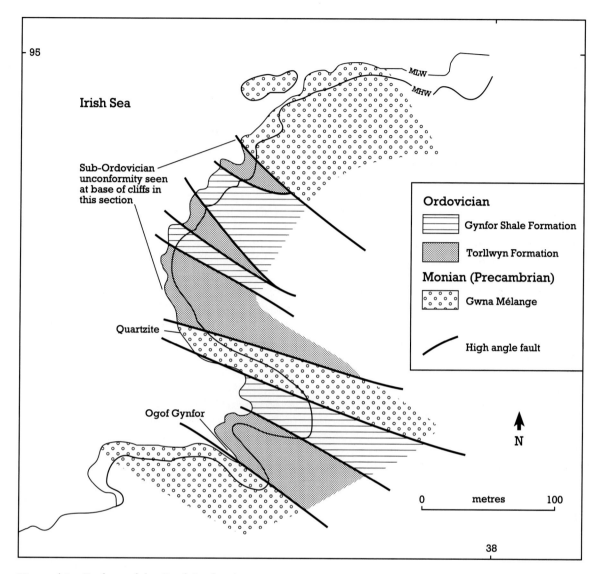

Figure 4.8 Geology of the Ogof Gynfor site.

Pebbles of Monian rocks occur in the Ordovician. The structure consists of two synclines and an intervening anticline, all faulted to some extent. The north limbs of both synclines are cut by reverse faults, which each give the appearance of footwall synclines. Thus the anticline becomes a hanging-wall anticline; the more northerly fault brings back the Gwna Mélange to the cliff top, and so no anticline is associated with it. The cleavage in the Ordovician rocks becomes more intense towards these faults. Lower-angle thrusting is also associated with the reverse fault at the northern end of the section.

Interpretation

The structural interest of this site lies in the relationships between the folding and faulting, particularly the way in which the faults are associated with folds and cleavage in the Ordovician sequence, and the way in which folds in the cover pass down into faults in the basement.

Prior to the work of Barber and Max (1979) all workers were agreed that the Gwna Mélange formed an integral part of the stratigraphical succession of the Precambrian Monian. Although Shackleton (1969) showed that the mélange was of sedimentary origin rather than tectonic, as maintained by Greenly (1919), both he and Bates (1972, 1974) agreed that the Ordovician deposition post-dated the Late Precambrian deformation of the Monian basement. Shackleton (1954, pp. 289, 291) particularly used these exposures to demonstrate the lack of décollement between the Precambrian basement and the Palaeozoic cover and the passage from clean-cut faults in the basement up into shear zones and folds in the cover. Barber and Max (1979), however, have proposed that much of the Monian of Anglesey has only suffered the same deformation history as the Ordovician above. For this reason and, in part, from palaeontological evidence (Muir *et al.*, 1979; Wood and Nicholls, 1973), they argue that the Gwna Mélange (part of their Cemlyn Unit) is of Cambrian age, and that the unconformity does not represent a significant tectonic or metamorphic event. Clearly, this hypothesis is of great significance to both the arguments concerning basement–cover relationships in the Caledonian Orogeny and also to the role of the Monian in the evolution of that orogeny.

The consensus among current research workers regarding the nature of this unconformity is unclear. Some recent publications on the evolution of Anglesey (for example, Gibbons, 1987) make no comment. No doubt, future research will be conducted on this important topic and this site will provide some of the crucial evidence. In particular, the continuity, or otherwise, of the cleavage between the units and their comparative metamorphic state will be important, as will be the deformational and metamorphic history of the Monian pebbles included in the Ordovician.

Conclusion

Ogof Gynfor provides important exposures of the unconformable contact between the Precambrian Mona Complex rocks and overlying Ordovician conglomerates and shales. Recent interpretations of the structure of Wales place great emphasis upon the significance of faults in the Precambrian basement and their influence, during the Caledonian Orogeny, on strain variation and structural style in the Lower Palaeozoic cover rocks. The structural significance of this site lies in the opportunity that it provides to examine the structural characteristics of and relationship between juxtaposed Precambrian basement and the Lower Palaeozoic, for instance, the way in which folds in the cover pass down into faults in the basement.

RHOSNEIGR (SH 317734)
R. Scott

Highlights

Rhosneigr provides a locality at which deformation on a variety of scales in the Ordovician cover sequence can be examined close to the underlying basement. The site provides unrivalled examples of strain variation around small-scale folds.

Introduction

The folded Ordovician greywacke sequence (the Nantannog Formation, of Arenig age) exposed at Rhosneigr is characterized by well-exposed minor folds. As minor folds can rarely be observed in the Lower Palaeozoic succession elsewhere in North Wales, Rhosneigr provides an important locality for their study. Folds of sandstone units are demonstrably non-cylindrical on a variety of scales. Excellent cleavage fans can be observed in the mudrock units surrounding the folded sandstone layers (Figure 4.9).

The Ordovician rocks of Anglesey were studied by Greenly (1919) who presented illustrations of the minor folds at Rhosneigr – see Figures 261 and 262 in Greenly (1919). It was not until Shackleton (1954) erected a general model for North Wales that the Lower Palaeozoic cover on Anglesey was reconsidered from a structural viewpoint. Whalley (1973) presented a thesis devoted entirely to a structural study of the Rhosneigr locality. A study of structures in the Lower Palaeozoic of Anglesey, including a description of Rhosneigr was undertaken by Bates (1974) who, like Shackleton (1954), emphasized the importance of basement control on deformation in the cover. The Rhosneigr locality was also mentioned in the paper of Barber and Max (1979).

The excellent degree of exposure at the site has allowed theoretical models of cleavage formation to be developed. In addition to the work of Whalley (1973), detailed studies include those of Knipe and White (1977) and White and Knipe (1978) which are of international significance in the study of cleavage formation. The site has also appeared in field guides (Barber *et al.*, 1981; Bates and Davies, 1981).

Description

The GCR site consists of an area of wave-cut

Figure 4.9 Rhosneigr, Anglesey. Tight minor folds in thin sandstones exemplify the intensity of the deformation in north-west Wales. The enclosing slates have been the subject of studies on the nature of slaty cleavage and strain variations around folded layers (penknife, centre, is 6 cm long). (Photo: J. Treagus.)

platform located to the west of the town. Some of the features of interest are illustrated by stereographic projection (Figure 4.10) and by line drawing (Figure 4.11).

An Ordovician greywacke sequence is deformed by upright folds on a variety of scales. Cleavage dips steeply to the north-west, as generally does bedding, but at moderate angles. The dominant sense of vergence is therefore toward the southeast. The greywacke sequence is dominated by shales, with folds delineated by discontinuous sandstone beds, generally <0.3 m thick. Knipe and White (1977) defined two types of fold based on scale:

1. macrofolds with a wavelength >10 m and
2. meso-(parasitic) folds with a mean wavelength ~0.5 m.

Two scales of non-cylindricity (fold axis curvature) can also be identified:

1. individual fold axes curve markedly within a

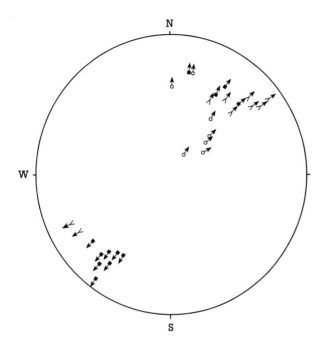

Figure 4.10 Equal-area stereographic projection of the plunge of minor fold axes at Rhosneigr. The site measurements are represented by the head of the arrow, and are divided into three subareas; circles = central, squares = NE and Vs = SW.

single exposure (for example, at SH 31637319) and

2. a general change in the plunge of folds is seen along the strike of the outcrop (Figure 4.10).

Individual mesofolds have divergent cleavage fans in the surrounding mudrocks while the cleavage refracts strongly through sandstones as convergent fans of series of spaced fractures perpendicular to bedding. Locally, cleavage in the shales is parallel to the outside arcs of the folded sandstones, defining a triangular zone of weak cleavage orientation, the finite neutral point of Ramsay and Huber (1987, p. 461). Figure 4.11 shows the pattern of cleavage displayed by the folds. A detailed account of the cleavage pattern is provided by Knipe and White (1977).

Exposed folded sandstone surfaces display fracture sets that are disposed symmetrically about fold hinges. At one location (SH 31637319), minor quartz slickensides on sandstone bedding surfaces suggest that some flexural slip was involved in fold development. Elsewhere, small quartz-filled fractures indicate extension in the outer arc of sandstone beds during folding.

Irregularly spaced rusty fractures can be observed in the mudrocks which appear to post-date the cleavage. They are generally subvertical

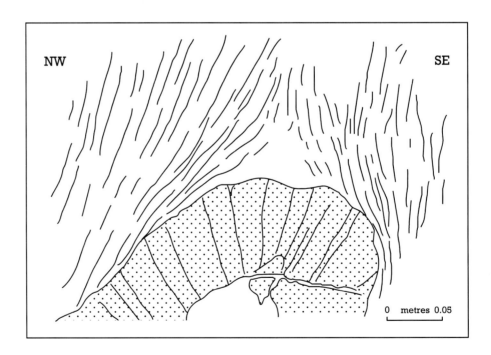

Figure 4.11 Rhosneigr. Line-drawing illustrating the strong cleavage refraction associated with the hinge of a meso-scale fold delineated by sandstone surrounded by pelite. A photograph of this fold appears in Ramsay and Huber (1983, figure 10.18).

and often strike approximately parallel to cleavage, but their orientation and spacing are very irregular. They occasionally reach widths of a centimetre and many contain a breccia of pelite fragments. At SH 31727338, folded sandstone beds are displaced along these rusty fractures, dominantly in a sinistral sense, by up to 1 m. A minor occurrence of *en échelon* quartz veining (at SH 31867350) is compatible with the sinistral displacement on the subvertical fractures. The fractures are also associated with small-scale thrusting of fold pairs (see Greenly, 1919; Figure 263).

Interpretation

The Ordovician of Anglesey lies unconformably on the Precambrian Mona Complex, many of the contacts being faulted. The majority of work on Anglesey has concentrated on the basement lithologies and their deformation. The general consensus is that this deformation was Precambrian in age (Roberts, 1979). However, as stated above, other workers have suggested that the Ordovician and Gwna Mélange were deformed together, for the first time, during the later stages of the Caledonian Orogeny (Barber and Max, 1979).

Shackleton (1953, 1954) stressed the importance of basement control on deformation of the Lower Palaeozoic cover in North Wales. He also emphasized the absence of a major décollement between cover and basement. The pattern of fold and cleavage orientations indicated to Bates (1974) that pre-existing structural trends in the Mona Complex dictated the deformation pattern in the Ordovician succession. However, the orientation of structures and degree of deformation at Rhosneigr are similar to those observed in similar lithologies on mainland North Wales where, presumably, the basement at depth had a much reduced influence on deformation in the cover.

In particular, it is interesting to note that, despite the close proximity (a few hundred metres) of the basement, the non-cylindrical nature of minor folds at Rhosneigr is compatible with the periclinal form of major folds in Snowdonia. The implication is clearly that, although the basement may have controlled the orientation of structures, the style and degree of deformation were controlled largely by processes acting within the cover. This is compatible with the 'thin-skinned' model of structural evolution (Coward and Siddans, 1979; Campbell *et al.*, 1985) and the strike-slip model of Woodcock (1984a).

Strike-slip or oblique-slip transpression during the Caledonian Orogeny may be used to explain the structures seen at Rhosneigr (see below), particularly considering its location close to faults on which major strike-slip motion has been proposed (Nutt and Smith, 1981). The rusty fractures on which sinistral displacement is apparent could with some certainty be ascribed to strike-slip movement, and a slight angle between axial traces and the strike of cleavage (Knipe and White, 1977) could be attributed to an earlier oblique-slip during the main deformation. However, considering the small-scale examples of basement faults passing up into the Ordovician cover sequence on Anglesey (see Ogof Gynfor above), perhaps the recently developed models of Smith (1988), Wilkinson (1988), and Wilkinson and Smith (1988) offer the most suitable explanation of the structure. Thus strike-slip movements on reactivated basement fractures may have provided, in a transpressional regime, the local heterogeneous strain manifested by the obliquity of cleavage to the axial planes and the late sinistral fractures.

Minor folds are rarely seen in the Lower Palaeozoic rocks of North Wales. The excellent exposure of the minor folds at Rhosneigr has made them suitable for detailed theoretical studies (Whalley, 1973; Knipe and White, 1977; White and Knipe, 1978). Knipe and White (1977) produced a detailed study of strain distribution in natural folds using a symmetrical meso-anticline from Rhosneigr, and later (White and Knipe, 1978) used slate from the site in the development of a model to explain cleavage initiation.

Conclusions

The Rhosneigr site, with its alternations of hard sandstones and soft mudstones, provides an excellent example of small-scale, non-cylindrical folding. This folding occurred during the Caledonian mountain-building phase, affecting a sequence of Ordovician sedimentary rocks. With its wealth of minor folds which are uncommon elsewhere in North Wales, the site provides an important locality at which the morphology of such folds can be compared with that of the major Caledonian folds. The outcrops here, very close to the underlying crystalline Precambrian basement, provide a location at which the influence of basement control on Caledonian deformation can be assessed, by comparison with localities on the

Welsh mainland where the cover was thicker and the structures that were developed were presumably further away from the influence of the ancient basement. Here the structures are similar to those seen further south in North Wales, and therefore are not thought to be greatly influenced by the Precambrian basement. This is an important issue in the interpretation of the Caledonian structure of Wales. The quality of these outcrops has enabled a detailed theoretical analysis to be undertaken on the basis of the relationship between cleavage and folding, and this has made a significant international contribution to studies of the origin of cleavage (Whalley, 1973; Knipe and White, 1977; White and Knipe, 1978).

CWM RHEIDOL
(SN 70057955–71147920)
W. R. Fitches

Highlights

This site illustrates the morphology and style of small-scale folds which are parasitic to major folds produced during the Caledonian Orogeny; folds on this scale are uncommon in most parts of the Welsh Basin. The site also exhibits the poor cleavage characteristic of west Central Wales, and provides examples of pencil cleavage.

Introduction

The Cwm Rheidol site has been chosen to represent a profile through a series of small-scale folds, developed on the limb of a major fold, the Plynlimon Dome, in a section of Lower Silurian sedimentary rocks. The structures at this site have not been described in the literature, although they resemble those discussed by Tremlett (1982), Craig (1985), and Cave and Hains (1986) in other parts of Central Wales. The Lower Silurian rocks of this region are described by Cave and Hains (1986).

Description

The 1 km-long track section exposes a 400 m-thick succession of well-bedded sandstones, siltstones and mudstones. The succession is the 'right way up' on the evidence of the cross-lamination and ripple-marks which abound in the section. The

sheet dip over the whole section is about 20° to the WNW, consistent with its position on the south-west flank of the Plynlimon Dome. The Silurian sedimentary sequence has here been shortened by 17–23%.

Folds occur on several scales, with wavelengths ranging from *c.* 400 m down to 0.10 m; most are in the range 1–10 m (Figure 4.12A). The folds are upward-facing, according to younging evidence, have upright NNE–SSW axial planes, plunge gently to moderately to the SSW, and are symmetrical or Z-folds in down-plunge profile. The variation in amount of plunge (horizontal to 40°SSW), obtained from stereograms of bedding (Figure 4.12B), of cleavage–bedding intersections (Figure 4.12D) and direct measurements of fold hinges, is probably due to non-cylindrical fold morphology, although there are indications of two distinct plunge populations.

The folds range from open to close, locally becoming tight with interlimb angles of less than 40°. Anticlines typically have rounded open profiles, whereas most of the synclines are close to tight with narrow hinge zones. This geometrical pattern resembles the cusp-and-lobe, or mullion structure described elsewhere by Sokoutis (1987) and Ramsay and Huber (1987, p. 397). On a smaller scale, cusp-and-lobe style structures are commonly developed in sandstone beds and are clearly seen on many bedding planes; wavelengths are usually in the 0.02–0.10 m range. These small structures, which have nucleated on sedimentary ripple structures in the sandstones in several instances, resemble the cusp-and-furrow described and illustrated by Cave and Hains (1986, Plate 17).

Most folds have Class 1B (parallel) geometry (Ramsay, 1967), but in some of the tighter anticlines and in most synclines the mudstone and siltstone layers have been slightly thickened in hinges to give a Class 1C geometry.

Cleavage is ubiquitous in the siltstones and

Figure 4.12 Cwm Rheidol. (A) Section along track showing bedding attitudes in siltstones and mudstones. Parts (B), (C) and (D) are equal-area stereographic projections of poles to bedding, poles to cleavage, and cleavage–bedding intersections respectively. (B) Dashed lines show great circle and small circle limits of the distribution and the large filled circle gives the pole to the great circle. (C) Dashed line represents mean cleavage attitude. (D) The two mean plunges of the cleavage–bedding intersections (open circles) can be seen to lie on the mean cleavage of (C) as does the pole to the bedding readings in (B).

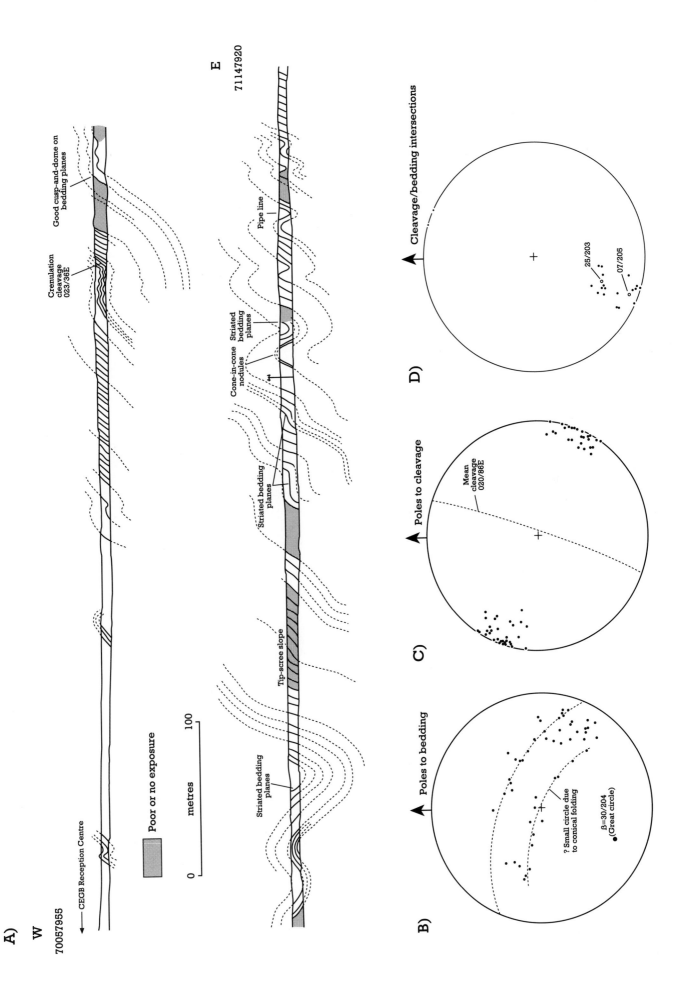

mudstones, but is uncommon in the sandstones. This fabric is seen at outcrop in fine-grained rocks as closely spaced (0.5–2 mm) surfaces which anastomose and are rough to smooth. Similar cleavage, elsewhere in the region, is seen under the microscope to comprise spaced surfaces along which pressure solution has taken place and very fine phyllosilicates are weakly aligned. The cleavage planes braid around large, detrital quartz grains and large chlorite–white mica stacks which are aligned in bedding. The cleavage in sandstones is poorly developed and widely spaced (more than 0.01 m in most instances).

The cleavage refracts strongly through layers of different ductility. It shows the fanning relationships to folds expected of layer-parallel buckle-folding processes. Statistically, from stereograms, the cleavage strikes 020° and dips 86°E (Figure 4.12C). Cleavage appears to be essentially axial planar to the folds – compare Figures 4.12B, C and D. A pencil cleavage, produced by the intersection between cleavage planes and the strong bedding fabric, is common in the section. Some 400 m from the western end of the site (Figure 4.12), a crenulation cleavage, oriented 023/36°E, overprints the main cleavage. Several bedding planes are striated by fine ridges and grooves which plunge WSW–WNW on westerly-dipping fold limbs, or ENE–ESE on easterly dipping limbs.

Interpretation

The folds in this section have orientations and symmetries consistent with their position in the south-west flank of the Plynlimon Dome, one of the major periclinal Caledonian folds of the Welsh Basin. The variation in plunge implies either non-cylindrical fold shapes or perhaps two populations of folds (Figure 4.12A and B). The cusp-and-lobe morphology observed at various scales is considered to be due to strong contrasts in ductility of the layers, the sandstones having the greater rigidity. The presence of large-scale structures with this form, the synclines being the cusps, may imply that the section is underlain by a thick rigid layer.

Measurements and calculations made from Figure 4.12A reveal that the amount of horizontal shortening accomplished by the folding ranges from 17.5% in the west and centre of the section to 28% in the east. These figures are similar to those obtained by Craig (1985) by measurements of distorted concretions on the Cardigan Bay coast.

The cleavage in the section is typical of the spaced anastomosing fabric which is widely developed in central Wales. Its microscopic characteristics have been described by Craig (1985), for example. The pencil cleavage, caused by intersection of bedding fissility and cleavage, is also found extensively in central Wales, as described by Craig (1985) from the Cardigan Bay coast. Cleavage in the Cwm Rheidol section is parallel with, or fans, with respect to the axial traces of the folds in steep surfaces, and stereograms reveal that the fold hinges lie in the cleavage; the hallmark of axial-planar cleavage (Figure 4.12A–C).

Conclusions

The Cwm Rheidol site provides an almost continuously exposed, 1 km-long profile through Silurian sedimentary rocks (around 440 million years old) that were folded and cleaved during the Caledonian Orogeny (around 400 million years ago). Continuous profiles of this length are exceptionally rare, so the site offers an unusual opportunity to examine the styles, sizes, and orientations of small folds, which themselves are not common in the Welsh Basin. The site shows evidence of at least two phases of Caledonian deformation, the main folding, with associated cleavage (fine, very closely spaced, parallel fractures), and a later wider-spaced cleavage which locally cuts the early set. Moreover, the quality and length of the exposures enables accurate determination of the amounts of crustal shortening responsible for folding; that is, by how much the Earth's crust was compressed and shortened by Caledonian earth movements with consequent vertical extension of the crust. It is very rarely possible elsewhere in the Welsh Basin to make such determinations. This is therefore a valuable cross-section through the south-western flank of the Plynlimon Dome, one of the major periclinal folds of Wales.

ALLT WEN (SN 57227877–57677969)
W. R. Fitches

Highlights

The Allt Wen site shows some of the wide variety of small-scale structures that characterize the northern part of the early Silurian Aberystwyth

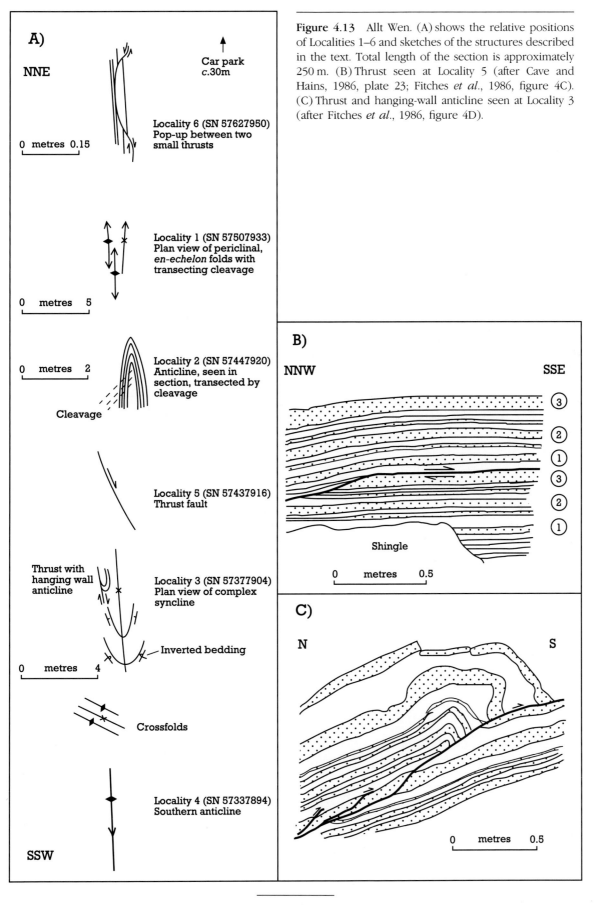

A)

NNE

Car park
c.30m

0 metres 0.15

Locality 6 (SN 57627950)
Pop-up between two
small thrusts

Locality 1 (SN 57507933)
Plan view of periclinal,
en-echelon folds with
transecting cleavage

0 metres 5

0 metres 2

Locality 2 (SN 57447920)
Anticline, seen in
section, transected by
cleavage

Cleavage

Locality 5 (SN 57437916)
Thrust fault

Thrust with
hanging wall
anticline

Locality 3 (SN 57377904)
Plan view of complex
syncline

Inverted bedding

0 metres 4

Crossfolds

Locality 4 (SN 57337894)
Southern anticline

SSW

B)

NNW   SSE

③
②
①
③
②
①

Shingle

0 metres 0.5

C)

N   S

0 metres 0.5

Figure 4.13 Allt Wen. (A) shows the relative positions of Localities 1–6 and sketches of the structures described in the text. Total length of the section is approximately 250 m. (B) Thrust seen at Locality 5 (after Cave and Hains, 1986, plate 23; Fitches *et al.*, 1986, figure 4C). (C) Thrust and hanging-wall anticline seen at Locality 3 (after Fitches *et al.*, 1986, figure 4D).

Grits Formation outcrop. Of particular interest here are folds which have unusually complex morphologies, the obliquity between the cleavage and some of the folds, and various brittle structures and veins.

Introduction

The 250 m-long coastal section comprises sandstones and shales of the Aberystwyth Grits Formation, a sequence of Llandovery-aged turbidites showing repeated Bouma B to E units (Wood and Smith, 1958; Cave and Hains, 1986). According to Price (1962) the section lies in the core of a major, Caledonian periclinal anticline; several folds with a 10–20 m wavelength, which are parasitic to this major fold, are exposed on the wave-cut platform. Thrusts, folds, and other structures within the site are illustrated and discussed by Craig (1985), Cave and Hains (1986), and Fitches *et al.* (1986) in the context of the timing of deformation with respect to the main Caledonian tectonism in the Welsh Basin.

Description

At the junction between the wave-cut platform and the coastal cliffs, are numerous small-scale structures of particular interest. Their positions are indicated on Figure 4.13A.

Folds

Periclinal anticline–syncline pair (Locality 1)

These folds have a wavelength of 2.5 m. The anticlinal axial plane is oriented 011/64°E, and its hinge plunges 10/020° in the north and 23/186° in the south. The structure is complicated in various ways:

1. The anticline comprises two *en échelon* anticlines, the one offset north and west of the other, without an intervening syncline;
2. In the syncline east of the anticline, a sandstone has been partly duplicated by a fault lying close to bedding;
3. A feeble cleavage in the anticlinal crest is oriented 036/64°E, and further down the axial plane appears to flatten. The cleavage is not axial planar to the folds but transects them in a clockwise sense.
4. The west limb of the anticline is disrupted by

an intermittently exposed composite structure that comprises a recumbent fold, ductile shear zone, and fault.

The recumbent fold is almost co-axial with the anticline, plunging 04/014°, but its axial plane is nearly horizontal (130/04°S). The recumbent fold was produced by ductile displacement along a westward-directed thrust which is marked in places by a 0.01–0.02 m-thick breccia. Cleavage is also deflected by this structure, implying that the fold–fault combination is late in the tectonic sequence.

Tight transected fold (Locality 2)

The crest of a tight, almost isoclinal fold, easily recognized by its 'gothic-arch' form, is exposed in the cliff-face. Its axial plane is N–S and upright, its plunge is nearly horizontal. The fold is upward-facing according to younging evidence in the west limb. However, the cleavage in that limb dips away from, and makes a large angle with, the axial plane so that the west limb of the fold is downward-facing with respect to cleavage; this unusual, non-axial planar relationship characterizes folds transected by cleavage.

Complex syncline (Locality 3)

The geometry of this fold has not been fully elucidated and it requires detailed grid-mapping. The northern part of the structure appears to be simple; its axial plane is N–S, upright, and the plunge is nearly horizontal. The southern part, however, closes on a highly curvilinear hinge which, from north to south, steepens from horizontal to vertical and beyond; this southern closure has the shape of the prow of an Indian canoe, which results in bedding being overturned. The southernmost end of this complex fold is hidden by shingle, but on the nearby wave-cut platform a series of small folds trend toward the syncline and are likely to have been responsible for refolding it.

Southern anticline (Locality 4)

This large, *c.* 10 m wavelength fold is an open to close, round-hinged structure with an upright, NNE–SSW axial plane and southerly plunging hinge line (*c.* 10/205°). Of particular interest is the crestal region in which the bedding planes are exposed. On the bedding there is a series of low-amplitude (*c.* 0.02 m), short-wavelength (*c.* 0.05 m) cuspate anticlines and rounded synclines (see Cave and Hains, 1986, Plate 17), which resemble the cusp-and-lobe fold mullions described by Sokoutis (1987) and Ramsay and Huber (1987).

On several bedding planes there are bedding-parallel ferroan dolomite, quartz, and chlorite veinlets up to 0.05 m thick. These veins are striated by slickensides which plunge approximately normal to the anticline hinge. In the cliff behind the fold crest several more of these veins are exposed, spaced at intervals of 0.05 to 0.50 m.

Faults

Particularly noteworthy is a series of small faults considered by Fitches *et al.* (1986) to be contemporary with the striated veins described above. Several examples are exposed along the section and three are described below.

Thrust fault (Locality 5)

A thrust fault with a minimum displacement of 8 m to the south or south-east causes a repeat of the turbidite beds at this locality (Figure 4.13B). The fault, in places marked by a centimetre-thick breccia layer and by fine quartz–carbonate veinlets, lies nearly parallel to the bedding in most places. In the central part of the exposure, however, the fault climbs a footwall ramp which is gently inclined and dips northward.

Thrust-hanging wall anticline

This structure, illustrated here in Figure 4.13C, lies in the western limb of the complex syncline described above. The thrust climbs a long, very gently inclined footwall ramp, which dips northward, and carries a hanging-wall anticline, in the southern limb of which the bedding is steep to overturned.

Opposing minor thrusts (Locality 6)

A thrust plane, which is mostly parallel with bedding, cuts up two opposite-dipping ramps to produce an inverted triangular fault block or 'pop-up' structure.

Interpretation

This site illustrates a variety of small-scale folds, relationships between cleavage and fold, faults and

veins. The folds have the complicated non-cylindrical, sometimes *en échelon*, morphologies to be found at several localities along the coast near Aberystwyth; they contrast with the relatively simple morphologies of the small folds occurring in most parts of the Welsh Basin. They have not yet been studied in detail and the causes of the complexities are uncertain. One possibility is that they result from accommodation space problems in the inner arcs of major folds, or alternatively, some of them at least were formed before the host strata were fully lithified. The site requires mapping and analysing in detail and the resulting information needs to be combined with that obtained at North Clarach and other coastal localities before a sound interpretation is possible.

The site's examples of folds transected by cleavage are important in arguments concerning the origin of this relationship. Craig (1985, 1987) accounted for the transection along the Cardigan Bay coastline, which includes Allt Wen, in terms of strike-slip deformation during the regional compression, along a major NNE–SSW zone, the Llangranog–Glandyfi Lineament.

The thrust faults, striated veins and some small folds have been interpreted by Fitches *et al.* (1986) as products of deformation before or during the earliest stages of the regional deformation. The bedding-parallel veins are regarded as products of hydraulic jacking. That is, high fluid pressures caused by impeded upward migration of fluids during burial of the sediment pile led to cavities being opened along bedding and minerals being deposited. It is suggested that the thrusts and displacements on the bedding-parallel veins are the result of gravity gliding, at some depth in the sediment pile, which took place before the onset of folding and cleavage development, but after lithification. Davies and Cave (1976) considered structures of this type to have developed before lithification because of their apparent association with dewatering structures in the sediments.

Allt Wen is an important site illustrating the complex morphologies of the small-scale folds that characterize the Aberystwyth part of the Welsh Basin. The origin of these folds is not yet understood and is the subject of ongoing research. The site also provides examples of rare cases of small folds transected by cleavage which is a topic under close scrutiny, not only in the Welsh Basin (Woodcock *et al.*, 1988), but in other parts of the British and North American Caledonides (Soper *et al.*, 1987); explanations of the phenomenon will lead to a clearer understanding of the plate tectonic evolution of the basin.

Several of the small-scale structures at Allt Wen have been illustrated and discussed in recent publications dealing with the timing of the structures, with respect to lithification and the regional deformation. These topics remain controversial and the site is likely to receive further attention by researchers.

Conclusions

The site at Allt Wen includes a whole suite of structures, folds, cleavage (very fine, closely spaced, parallel fractures), and faults, which affect the early Silurian-aged Aberystwyth Grits. These structures are the result of extreme compression during the Caledonian mountain-building episode, around 400 million years before the present. The complexities of the thrusts (low-angle faults), folds, and cleavage here have yet to be studied and explained fully. For instance, the cleavage slightly cuts across the planes that bisect fold limb-pairs (the axial planes). This is an uncommon relationship in fold belts but, by analogy with other Caledonian terranes in Britain, may be related to the oblique approach of the colliding continents as the Iapetus Ocean closed.

Some structures here are thought to have been generated before the main (Devonian) deformation phase of the orogeny, and to have been formed as contortions in the perhaps still wet sediment pile, or perhaps as the sediments moved downslope under the influence of gravity. Upon this folding would have been superimposed the regional tectonic pattern of folding and cleavage. This remains a site with much potential for future study.

NORTH CLARACH
(SN 58508410–58578446)
W. R. Fitches

Highlights

The folds in the Llandovery Series (Lower Silurian) sedimentary rocks at the North Clarach site have unusually complicated geometrical relationships with the cleavage, which makes the site particularly important for research into the relative timing of these two types of structure. The site also provides clear, small-scale examples of cleavage transection,

a phenomenon of topical research interest in the Welsh Basin. This phenomenon has implications for the understanding of the plate tectonic setting of the basin.

Introduction

This site consists of a wave-cut platform, submerged at mid- and high-tide, showing folded and cleaved turbiditic sandstones, siltstones, and mudstones of the Aberystwyth Grits Formation. These sedimentary rocks have been described by Wood and Smith (1958), and Cave and Hains (1986). The principal points of interest here are the examples of repeated (or progressive) folding and of various geometrical relationships between folding and cleavage. Cleavage transects some folds, is axial-planar to others, and locally is itself apparently folded. The platform at Clarach has been mapped at a scale of 1:50 by Mrs R. Johnson, formerly of University College of Wales, Aberystwyth. Her map is reproduced in simplified form as Figure 4.14. The structures displayed in the platform were briefly commented on by Fitches and Johnson (1978).

Description

A clearly recognizable example of a fold transected by cleavage is found at the northern end of the site (Locality 1, Figure 4.14). The fold is a periclinal anticline with an upright, N–S axial plane and hinge line which plunges gently north and south. Cleavage, which is well defined in the shale layers, strikes obliquely to the axial plane in a clockwise sense and dips to the WNW at a lower angle.

Several other examples of folds transected by cleavage are exposed on the platform, notably at Locality 2 where cleavage cuts both limbs of an open anticline on the western limb of a larger syncline.

Around Locality 3, where the rocks are conspicuously cut up by a complex of small faults, many of which are eroded out as gullies, several small folds appear to deform the cleavage. These folds, with metre-scale wavelengths, are close to tight structures on variable, but mostly N–S,

Figure 4.14 North Clarach. Fold–cleavage–fault relationships on wave-cut platform (modified from map produced by R. Johnson, University College of Wales, Aberystwyth, 1977). Localities 1–4 referred to in the text.

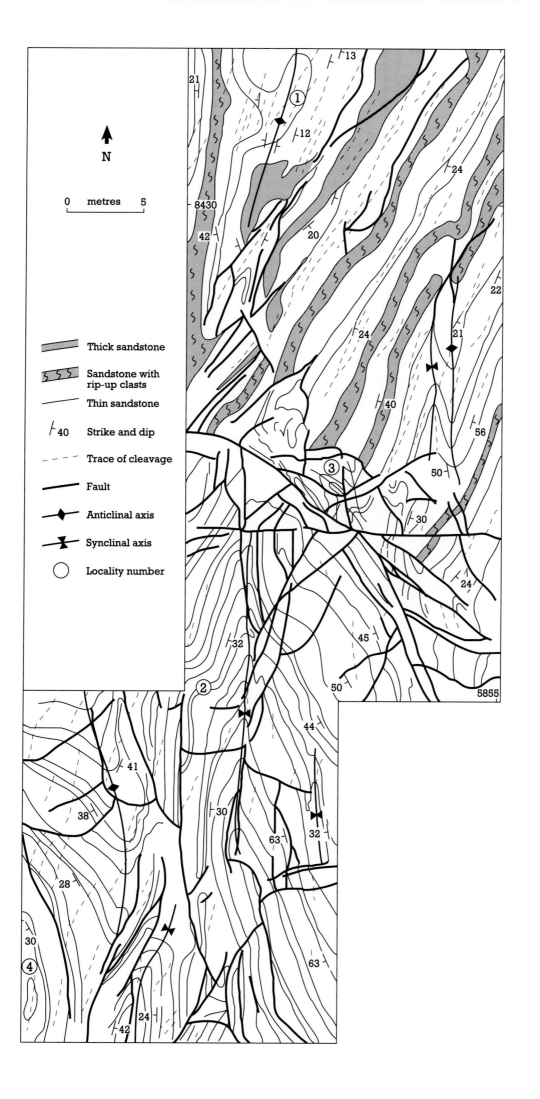

upright axial planes. They are unusual for their very steep plunge. Cleavage in shale beds is nearly parallel to the bedding and appears to pass round the fold hinges. The precise disposition and pattern of cleavage throughout a particular fold has not been determined because of the feeble, often ill-defined nature of the cleavage, the difficulties of distinguishing it from compactional bedding-plane fabrics, and the problems of sampling these fissile rocks for sectioning.

Two stages of folding can be demonstrated around locality 4 where, immediately north of a conspicuous gully, an open anticline trends N–S. To its west, a very gentle, saucer-like periclinal syncline trends approximately NNW–SSE. The syncline has been superimposed on the anticline, causing a gentle plunge depression in the anticlinal hinge. The cleavage in this area bears no simple geometrical relationship to either of the folds, and it appears to transect both structures.

Interpretation

The reasons for the variable orientations of the folds, the superimposition of folds on each other, and the different geometrical relationships between folds and cleavage are subjects of ongoing investigation and are not yet completely understood. One interpretation of all these phenomena is that the folds and cleavage, and possibly some of the small faults as well, are due to accommodation in the core of a major anticline. In this case the variable attitudes and overprinting relationships of the folds, and the incongruent relationships between cleavage and folds could be explained by complex stress reorientations during progressive tightening of the host structure.

A second interpretation is that the structures were caused by progressive deformation accompanying transpression along the N–S Llangranog–Glandyfi Lineament, the central part of which is likely to pass through the area a short distance inland from North Clarach. Craig (1985, 1987) has described how, further south in the lineament, folds developed at different times during displacement along the zone; refolding, as seen at North Clarach, can be produced as early formed folds rotate to a new alignment with respect to stress axes in the transpression zone. Similarly, cleavage does not necessarily develop contemporaneously with folds in transpression zones. It can precede, follow, or form at the same time as folds, leading to the various types of geometrical relationships

(superimposed folds, transecting cleavage) observed at North Clarach.

A third interpretation of the apparently disorganized geometrical relationships between folds and between folds and cleavage, based on studies elsewhere, is that these structures developed in unlithified or only partly lithified sediments, which might have produced initial irregularities of bedding planes.

Conclusions

The North Clarach site contains examples of structures which are important in understanding the sequence of events and processes which characterize the Caledonian mountain-building episode in this region. Here are seen folds which have been refolded, and also complicated relationships between folds and cleavage. Contrasting relationships are seen: folds are present with associated cleavage (very fine, closely spaced, parallel fractures), which parallels the planes which bisect fold limb-pairs (that is, the cleavage is axial planar). In other examples, the cleavage cuts across (transects) the fold axial plane, whereas in other situations folds are seen to deform, and therefore apparently post-date, the cleavage. These relationships have been explained in various ways: as the product of the progressive tightening of the major fold in which the site lies, as a product of progressive adjustment in relation to a major structural lineament nearby, and finally, as being due to various irregularities in the sedimentary rock pile that were present before deformation commenced.

This site is important and it is likely to yield important evidence on folding and cleavage-forming processes, the timing of deformation with respect to lithification of the host sediments, and on the regional structure of west Central Wales.

CORMORANT ROCK (CRAIG Y FULFRAN) (SN 583830)
W. R. Fitches

Highlights

The Cormorant Rock site exposes a series of folds produced during the Caledonian Orogeny. The cleavage has unusually complex geometrical relationships with the folds. Small-scale folds, termed 'tectonic ripples' in the older literature, have

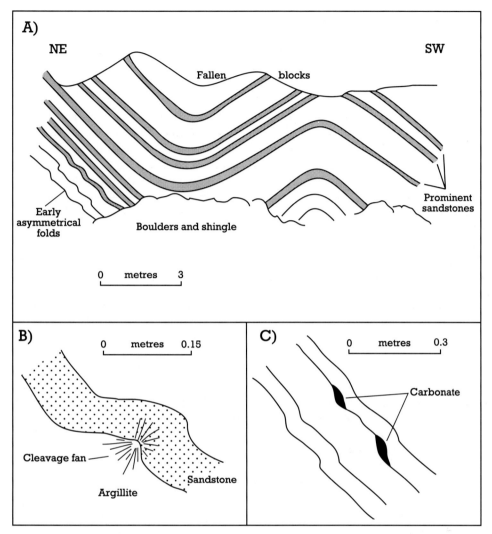

Figure 4.15 Craig y Fulfran. (A) Regional deformation folds with early asymmetrical small-scale folds on the northeastern limb, further illustrated in (B) and (C). (B) shows cleavage fans and (C) saddle-reefs in hinge zones (after Fitches *et al.*, 1986, figures 6(A), (C), and (B)).

recently been interpreted as products of deformation which preceded the main Caledonian tectonism.

Introduction

The site is located at the foot of the main sea-cliffs opposite the Cormorant Rock sea-stack. The rocks are interbedded turbidite sandstones, siltstones, and mudstones of the Llandovery Series (Lower Silurian) Aberystwyth Grits Formation (Wood and Smith, 1958; Cave and Hains, 1986). The site has been selected for the unusually complex geometrical relationships between cleavage and folds which it shows, and for its anomalous small-scale

folds, first recorded and described as 'tectonic ripples' by Wood (1958) and recently discussed by Fitches *et al.* (1986).

Description

The dominant structure comprises an anticline–syncline pair (Figure 4.15A) with a wavelength of *c.* 10 m, steep and nearly N–S axial planes, and a gentle to moderate southerly plunge. These folds probably represent regional deformation structures. Cleavage is mostly weak and ill-defined, and has complex relationships with the folds. In places it is axial planar, but in the anticlinal hinge zone it forms a very open downward-divergent fan and lies at only a small angle to bedding.

Of particular interest are the small-scale folds of bedding developed in the eastern limb of the syncline. These 'tectonic ripples' (Wood, 1958) are illustrated in Figure 4.15B and C. These folds are open to close, S-shaped asymmetrical structures with wavelengths of about 0.30 m and amplitudes of about 0.10 m. Their axial planes are reclined and hinge lines plunge gently northward; they are not coaxial with the host syncline. A weak cleavage, marked by feeble grain alignment in sandstone and a fissility in finer rocks, is restricted to hinge zones, where it makes upward convergent fans in shales and downward convergent fans in sandstones (Figure 4.15B). Calcite-filled saddle-reefs have developed in some inner arcs of these small folds, at the interfaces between sandstone and shale (Figure 4.15C).

Interpretation

The anticline and syncline at this site are considered by Fitches *et al.* (1986) to represent regional, end-Caledonian folds on the basis of their N–S, steep axial planes which are parallel with other regional folds of the district. The cleavage is probably also an end-Caledonian structure, but the reason for its very strong divergence in the anticline hinge is unclear. One explanation is that it formed in the site of intense extensional strain on the outer arc of the folds, contemporaneously with folding, as illustrated in a theoretical model by Ramsay and Huber (1987, Figure 21–26), for example, and from Snowdonia by Wilkinson (1988, Figure 4.31). Alternatively, the cleavage may locally have formed after the host layers were folded, as observed at the North Clarach site.

The 'tectonic ripples' in the limb of the syncline are not parasitic to that larger fold; they have the wrong sense of asymmetry, their axial planes and hinge lines are incongruent with respect to those of the larger fold, and the main cleavage appears to cut obliquely across them. Fitches *et al.* (1986) inferred that they preceded the development of the syncline and attributed them to gravity gliding toward the north-west which took place before, or as a precursor to, the regional deformation. On these grounds, the small folds were assigned to a family of early structures, other members of which are considered to be represented at Allt

Wen and Traeth Penbryn, for example.

Recent examination by the author of other 'tectonic ripples' in the Aberystwyth Grits between Clarach and Borth, a few kilometres north of Cormorant Rock, has prompted consideration of an alternative explanation for these structures. Rather than being folds caused by compression along bedding, the ripples are more likely to be narrow, almost uniformly spaced ductile shear zones. Each shear zone is inclined at a high angle to bedding which it deflects down to the west by a few centimetres. A weak cleavage, developed along each zone, is oriented obliquely to the zone margin in a manner consistent with downwards ductile displacement on the west side. These structures are almost certainly older than the regional folding and cleavage for reasons discussed above.

The presence of the calcite-filled saddle-reefs in the hinges of some 'tectonic ripples' implies that the host sediments were sufficiently cohesive to allow the opening of cavities during the early deformation. Such brittle behaviour may imply that the sediments were at least partially lithified by this stage.

Conclusions

The small folds of the Cormorant Rock site provide information on deformation that preceded the main Caledonian folding and cleavage development in this part of the Welsh Basin. They have formed either by gravity gliding or as a result of movement on small slip surfaces (ductile shear zones). The larger-scale, but still comparatively lesser-order folding, and cleavage, here were the product of the Caledonian mountain-building phase. They are orientated in the typical N–S manner of other Caledonian structures in the region, but the cleavage is locally stongly divergent from the planes which bisect the fold limb-pairs (axial planes) of the folds. It is not clear why this is the case, but it may be due to some localized strain in the outer arcs of the folds or it may, more simply, indicate that the cleavage was formed after the folding. The site is an important one for studying the effects of the Caledonian Orogeny in this region. Because the site exemplifies relationships between folds and cleavage which are not yet fully understood, it is an important locality for current and future research.

PONTERWYD QUARRY (SN 74028085)
W. R. Fitches

Highlights

The quarry near Ponterwyd has been selected to show examples of the rare, small-scale folds and cleavage which are superimposed on the regional Caledonian structures of west Central Wales. The site also provides an example, possibly unique, of the relationship between lead veining and the later deformation which is assumed to be Caledonian.

Introduction

This site exposes Lower Silurian interbedded fine sandstones, siltstones, and mudstones, described by Cave and Hains (1986), which are located in the south-western part of the Plynlimon Dome. Of particular interest are the folds and cleavage which are imposed on the main cleavage. These late structures, together with a vein of galena which cuts some of them, are described, discussed, and illustrated by Fitches (1972). Fitches (in discussion of Phillips, 1972) drew attention to their relevance to the timing of mineralization with respect to deformation of the rocks of the Welsh Basin.

Description

Figure 4.16A, a sketch plan of the quarry, gives the location of the localities discussed below. The bedding generally strikes NNE-SSW and dips steeply westward, and is the right way-up according to the abundant cross-lamination, ripple marks and trace fossils. Cleavage strikes NNE–SSW and dips very steeply westward.
4.16A, B & C

In the west face of the quarry, immediately north of the spoil tips (Locality 1, Figure 4.16A), there are several recumbent folds of the cleavage and bedding. The axial plane of one fold is oriented 360/12°W, the hinge plunging 20/191°. The folds have open to gentle profiles, wavelengths of *c.* 0.20 m and amplitudes of *c.* 0.05 m; hinge zones are narrow and limbs are planar, giving a chevron style. A feeble crenulation cleavage is axial planar to the folds. This fabric is replaced locally by zones of very thin quartz veinlets which are also axial planar.

Several more late folds, forming a conjugate set, are exposed in the east face of the quarry, in the north-east corner (Locality 2, Figure 4.16A and 4.14B). Wavelengths are in the range 0.30 m to 2 m. Some of the folds have axial planes striking NNE–SSW (010–015°) and dipping moderately to the east (45–50°), and hinge lines that plunge gently to moderately northward 15–30° to 010–020°. Other folds in the conjugate set have axial planes oriented approximately 035/20°SE, and hinge lines plunging *c.* 05/205°. A feeble crenulation cleavage is axial planar to the steeper folds. Bedding surfaces in this fold complex are commonly slickensided (striations and quartz slickencrysts) as a result of flexural slip during folding.

Cutting folds of both orientations is a 0.15 m wide zone of quartz veins (064/84°SE) which is exposed from the quarry floor to the top of the east wall. One 0.03 m-wide vein in the middle of the zone is composed of galena.

The north face of the quarry (Locality 3, Figure 4.16A and C) exposes the profiles of the recumbent and inclined folds which make up the conjugate set of late folds described above. The folds at the eastern end of this face are those at Locality 2. One of these folds is of particular interest for the mineralized veinlets associated with it (see Figure 2 of Fitches, 1972). Part of its hinge zone is occupied by a carbonate-filled saddle-reef, and its upper limb contains a tension gash array, formed during folding, in which quartz and pyrite (and possibly chalcopyrite although it is too small to identify in the field with confidence) have segregated.

Interpretation

Folds and crenulation cleavages which are imposed on the main end-Caledonian folds and cleavage have been reported from various parts of the Welsh Basin (Roberts, 1979; Martin *et al.*, 1981; Fitches, 1972; Smith, 1988). Fitches and Roberts considered that the late (post-main deformation) folds with flat-lying axial planes, like many of those exposed in the Ponterwyd Quarry, belong to a regional set, while steep folds imposed on the recumbent ones represent a younger regional set. It was pointed out by Tremlett (1982) and Craig (1985), however, that these locally developed, late structures commonly appear to be associated with faults, and that they are therefore unlikely to be products of regional events. This fault-related explanation is supported by observations in other parts of the Welsh Basin. Near the Bala Fault, kink

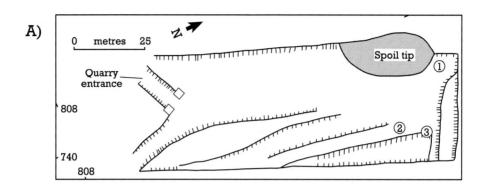

A)

B)

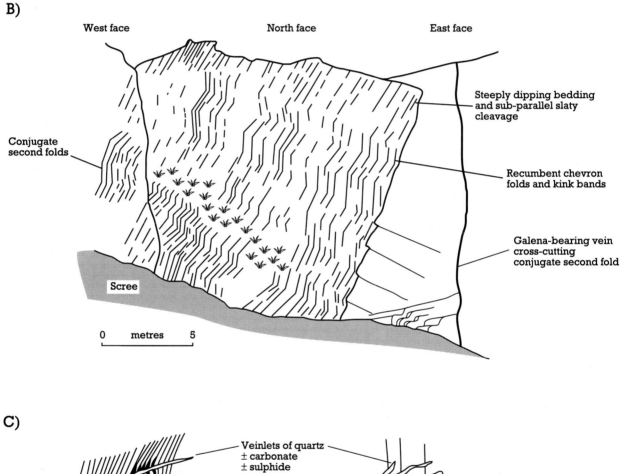

West face   North face   East face

Steeply dipping bedding
and sub-parallel slaty
cleavage

Conjugate
second folds

Recumbent chevron
folds and kink bands

Galena-bearing vein
cross-cutting
conjugate second fold

Scree

0   metres   5

C)

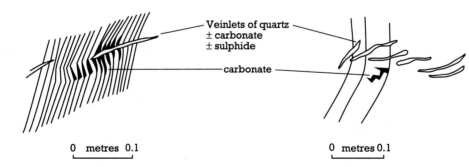

Veinlets of quartz
± carbonate
± sulphide

carbonate

0   metres 0.1   0   metres 0.1

Figure 4.16   Ponterwyd Quarry. (A) Plan of quarry showing the three localities described in the text. (B) Locality 2; steeply dipping bedding and subparallel slaty cleavage deformed by recumbent second-phase chevron folds and kink bands (after Fitches, 1978, figure 1). (C) Locality 3; saddle-reefs and tension gashes (after Fitches, 1972, figure 2).

bands and crenulations, for example, imposed on the main cleavage are spatially related with the fault zone (Bracegirdle, 1974; Fitches and Campbell, 1987).

The sulphide-bearing veins in the Ponterwyd Quarry provide information on the timing of mineralization with respect to deformation in this part of the Welsh Basin. The fact that pyrite, and possibly chalcopyrite, are found in tension gashes produced during the late folding indicates local segregation of sulphides from the host rocks into low-pressure regions during deformation. The galena vein, however, cuts across, and is therefore at least slightly younger than all the late structures. This relationship of galena mineralization to the late structures shows that in this case mineralization is unlikely to be related to dewatering of the sedimentary pile during the main end-Caledonian (late Silurian–early Devonian) deformation (Fitches, in discussion of Phillips, 1972).

Conclusions

Ponterwyd Quarry contains numerous, particularly well-exposed examples of folds imposed on, and therefore younger than, the main Caledonian structures. At this site these consist of cleavage (very fine, closely spaced, parallel fractures), and steeply dipping bedding in these Silurian strata. Superimposed on these two sets of features are a variety of folds (Z-shaped and S-shaped, and sometimes chevron-shaped) and a second generation of cleavage with parallel quartz veins. The superimposed folds and the cleavage are regarded as late-Caledonian structures found uncommonly in several parts of the Welsh Basin. Mineralized veins in the quarry are associated with the second generation of structures: they provide information, possibly unique in the Welsh Basin, on the timing of relationships between the formation of lead veins and the tectonic structures.

TRAETH PENBRYN
(SN 28755210–29095232)
W. R. Fitches

Highlights

The site contains the best-known examples of veins and hydraulic fracture breccias produced by brittle deformation, caused by high fluid pressures before, or during, the earliest stages of regional Caledonian tectonism.

Introduction

This site, a sea-cliff section, comprises folded and cleaved Upper Ordovician sedimentary rocks of the Tresaith Formation (Craig, 1985). These are cut by a series of veins and hydraulic fracture breccias which have been described, discussed and illustrated by Craig (1985) and Fitches *et al.* (1986). These structures are considered to belong to a family of early brittle structures, other members of which are represented at the Allt Wen, Cormorant Rock, and Llangollen sites.

Description

Craig (1985) and Fitches *et al.* (1986) have shown that the brittle structures affecting the rocks here preceded the folds and cleavage; they attributed them to dewatering processes before, or during, the onset of regional deformation.

The country rocks of the Tresaith Formation (Ashgill Series) are thinly bedded, fine sandstones, siltstones, and mudstones. The structure of the section is mostly uncomplicated bedding (065/20°SE) and cleavage (075/35°SE) maintaining nearly uniform orientations except for local interruptions by small folds, to which cleavage is axial planar.

The veins fall into two main categories:

1. hydraulic vein breccias, and
2. simple and *en échelon* veins:

1. Hydraulic vein breccias

The largest vein breccia (Fitches *et al.*, 1986, Figure 4B) is more than 10 m in length, 0.20 m in thickness, and is oriented about 030/70°SE. The vein minerals are predominantly quartz, with a little carbonate and traces of pyrite. The breccia fragments are angular pieces of local country rock, many of which are totally suspended in the vein minerals, whereas others remain partly attached to the vein walls and have been only slightly rotated and detached from the country rock. Craig (1985) and Fitches *et al.* (1986) showed that vein breccias were formed after the country rocks had undergone extensive diagenesis, but before the development of the regional cleavage.

2. Simple and en échelon veins

Veins composed chiefly of quartz, with minor carbonate and pyrite, are common in this section, typically about 0.05 m thick, steeply dipping and aligned mostly NNE, but varying widely in orientation. Most veins occur singly, but locally, notably near the large vein breccia, several veins are disposed *en échelon* in tension gash arrays.

Some of these veins cut the vein breccia. In places, they are deformed by small open to tight folds, with wavelengths of a few centimetres or tens of centimetres, to which the cleavage has an axial-planar or fanning relationship.

Interpretation

Microfabric evidence, obtained from breccia clasts in the vein breccias shows that the growth of phyllosilicates in the bedding compaction fabric (Craig *et al.*, 1982) preceded the formation of a pressure solution fabric, which is itself misaligned due to variable amounts of rotation of the host fragments. This evidence indicates that breccia formation followed diagenesis. That brecciation preceded the cleavage is shown in two ways. Firstly, the breccia is cut by later veins which were themselves folded and cleaved by the regional deformation. Secondly, under the microscope, an earlier grain alignment fabric in the breccia fragments is crenulated on planes which are parallel with the cleavage in the country rocks.

The simple and *en échelon* veins also preceded the development of cleavage because they are deformed by folds to which the cleavage is axial planar. Craig (1985), by measuring the lengths of veins around folds, calculated that the former have been shortened in the cleavage by about 30%.

The brecciation and deposition of the minerals hosting the fragments are attributed, by Craig (1985) and Fitches *et al.* (1986), to hydraulic fracture processes similar to those advocated by Phillips (1972) to explain the post-tectonic veins of Mid-Wales.

Craig (1985) and Fitches *et al.* (1986) considered that the vein breccias and other veins were caused by high fluid pressures which developed during burial, but after lithification, of the (Upper Ordovician) sediment pile. The vein breccias were interpreted as a manifestation of extension of the sediment sheet caused by gliding down a slope under gravity, and the *en échelon* veins were taken to represent the flank of a sheet subjected to strike-slip displacements. The various veins belong to a family of early structures, of which other members, mostly the structures of the toes and central parts of glide sheets, are represented elsewhere.

Conclusions

The Traeth Penbryn site contains various types of veins (vein breccias, simple and *en échelon* north-easterly striking veins), which have been shown to have preceded the folds and cleavage produced during the main phase of the Caledonian Orogeny. These veins are regarded as indicators of pre-tectonic or early tectonic deformation processes which are very rarely represented in Lower Palaeozoic strata. The vein breccias (veins containing angular rock fragments set in a matrix of the mineral quartz) are perhaps unique in the Welsh Basin context. They are thought to have been produced by the action of fluids under high pressure acting on fractures brought about by stretching, and perhaps sliding on a large scale, of the Ordovician sediment pile. Subsequently the whole area was affected by the Caledonian earth movements during Silurian to Devonian times; the simple veins (which are later than the vein breccias because they cut them) were cleaved and folded. The site provides the best exposures in Wales of veins and other structures which can be proved to pre-date the Caledonian mountain-building episode.

CA'ER-HAFOD QUARRY, LLANGOLLEN (SJ 215476)
R. Nicholson

Highlights

Carbonate veins here are platy and lineated. They lie along bedding and record movements, apparently tectonic, prior to the main Caledonian phase. These phenomena appear to be restricted to rocks of mid- to late-Silurian age.

Introduction

The Ca'er-hafod Quarry provides rare outcrops of a very distinctive suite of laminated carbonate veins, apparently lying along bedding, and inscribed with a very pronounced rectilinear, ridge-and-groove lineation. Both the mineral fabric of the veins and this lineation are older than the deformation episode which folded the Wenlock Series country rocks. The veins (spar beds or 'rhesog') here had their macroscopic features first described by Wedd *et al.* (1927), in the Geological Survey Memoir dealing with the Wrexham district.

Figure 4.17 Ca'er-hafod. Part of the quarry showing bedding dipping steeply south, and cleavage gently north. The outcrop of the central vein (top left to centre) shows minor folds plunging towards the observer, and ridge-and-groove lineation almost at right-angles to this. View looking east. (Photo: R. Nicholson.)

Wedd *et al.* noted the bedding-parallel nature of the veins, that they are affected by folding, the presence of a groove-like lineation on the veins, and the likelihood that there had been considerable amounts of movement along them. These authors also made clear the apparent restriction of the veins, in the Wrexham district, to the Wenlock Pen-y-glog Formation.

This slate formation, once worked on both north and south limbs of the gently easterly plunging Llangollen Syncline, has its subcrop marked by a string of disused quarries. They provide the only access to these carbonate veins, which were unknown in natural outcrops. The quarries also provide the most convenient areas in which to examine the regional cleavage; this has a moderate dip to the north, atypical of Caledonian North Wales. The Ca'er-hafod Quarry is on the north limb of the Llangollen Syncline (Figure 4.17).

The microscopic character of the veins was first described by Nettle (1964). Veins with the same mineralogy, fabric, and structure as those of the Llangollen Syncline, are found in the Middle Wenlock to Lower Ludlow rocks of areas east of Llanwrst, in northern Clwyd (Warren *et al.*, 1970). These authors, however, broadly link formation and deformation of the veins, with formation of the regional cleavage.

Description

Cleavage and bedding here have the characteristic E–W trend of this north-eastern section of the Welsh Basin. The disused quarry situated near Ca'er-hafod (Wedd *et al.*, 1927, p. 97; Nicholson, 1966, 1970, 1978) was referred to by Fitches *et al.* (1986, Figure 7D) by the name of the nearest house, Pont Glas. The 180 m-long working, opens to the east, is nowhere wider than 40 m, and is driven into the silty mudstones of the Pen-y-glog Formation. Its steep north and south walls are parallel to the strike of bedding (Figure 4.18).

Situated in the complexely folded, north limb of the Llangollen Syncline, bedding surfaces dip steeply to the south (e.g. 105/62°S). The accompanying cleavage dips moderately to the north

S N

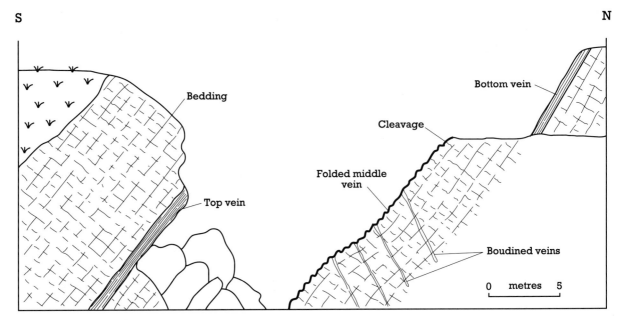

Figure 4.18 View looking west at Ca'er-hafod Quarry (Llangollen) showing steeply dipping Wenlock country rocks and spar beds (veins).

(e.g. 093/55°N), bedding and cleavage here being about perpendicular to one another, suggesting that the rocks of the quarry lie near a fold hinge. Although the Llangollen Syncline, like many of the subsidiary folds of its north limb, plunges gently to the east, the folds affecting the three veins, and the bedding–cleavage intersections in the enclosing slate, all plunge gently westwards. The moderate northerly dip of the cleavage, is typical of the syncline as a whole.

Three principal laminated veins occur in the quarry (Nicholson, 1978). All of them have a well-developed ridge-and-groove lineation on the calcite laminae of which they are made. The bottom vein, forming the north wall of the quarry, shows how this lineation may change substantially in pitch, even from the surface of one-millimetre-thick lamina to the next. The top vein, which lies at the base of the south wall of the quarry, is approximately 0.10 m thick, the thickest of the three. The central vein is the thinnest and most thoroughly laminated. As a result, it shows more regular and intensively developed folds than the other veins (Figure 4.17).

The bottom vein is partly composed of a cemented breccia. Both in this and other respects, it closely resembles that exposed in the north wall of the easternmost of the quarries of Moel y Faen (SJ 18874772; Wedd *et al.*, 1927). The other two veins of the Ca'er-hafod Quarry do not contain breccia zones.

The central vein, about 10 mm thick, is exposed for tens of metres on a mesoscopic, folded bedding surface (Figure 4.18). The published description of the modifying effects of this deformation on the primary fabric of generally vein-parallel, single-crystal plates of calcite is based on material from this central vein (Nicholson, 1978; microprobe data on calcite composition, Hamdi Lemnouar, 1988). The very numerous, small folds are periclinal in form, each having a hinge length of some 0.20 m. These hinges lie parallel to those of the larger folds. They all plunge parallel to the bedding–cleavage inter-section on the folded surface and have their axial surfaces parallel to the cleavage of the slate in which they are embedded. All these field relation-ships confirm the interpretation based on micro-scopic evidence, that folding was part of the regional deformation, taking place after the veins were already fully formed.

The slates of the Ca'er-hafod Quarry also contain discordant calcite–quartz veins, apparently linked in origin with the laminated, bed-parallel veins. Instead of being folded during regional deformation, they were boudined. The early formation of these discordant fractures may be used as an indication that the cleavage here is not a structure formed during burial and related compaction and water loss (Davies and Cave, 1976). This has importance for the assessment of proposals made concerning the place of vein systems in the regional tectonics of Caledonian Wales. Fitches *et al.* (1986), for example, making passing reference to the Silurian rocks of the Wrexham district, suggest that such veins may have formed as water driven from sediments during burial was injected along chosen bedding horizons. It is supposed that such injection allows the detachment of upper levels of the sediment column, freeing them for the lateral movement that the formation of the lineation on the veins requires.

The wide, folded, bedding surface on which the central vein is exposed, also reveals the way in which a number of highly discordant veins are joined to the central vein, in a direction subparallel to the bedding–cleavage intersection. These sheets, at most 10 mm in thickness, dip more steeply to the north than cleavage, and have been extended in their plane, boudined, during the regional deformation. The evidence of extension is found in the repeated quartz-filled zones, less than a millimetre thick, that cross discordant sheets, in directions sub-perpendicular to them. The patterns of fractured calcite crystals on either side of these zones match, although they are separated by the fibrous quartz that now fills the zones (Nicholson, 1966). The formation of these extension structures, developed in planar bodies lying at low angles to cleavage when they were deformed, is consistent with the simultaneous formation of folds in the veins, lying about at right-angles to cleavage.

In all three veins, the laminae are made up of vein-parallel, single-crystal, calcite plates, which have their crystallographic c-axes at right angles to their planar surfaces. Plates very commonly have thickness to lateral extent ratios of at least 100. Their thicknesses range from 0.1 mm to over 10 mm. The amplitude of the folds later imposed on the fabric varies in direct proportion to the vein thicknesses. Deformed plates are distinguished by complex developments of *e*-lamellae (Nicholson, 1966). The calcite crystals of the discordant, boudined veins, are also plate-like in shape, although arranged in various orientations, oblique to vein walls. These plates are separated from one another by relatively coarse-grained quartz.

Nettle (1964) attributed the unusual laminated structure of the veins to the modification of earlier fabrics during regional deformation and metamorphism. This interpretation was challenged by Nicholson (1966), who emphasized the primary nature of their laminated structure, and the way that it was affected distinctively by the succeeding episode of deformation, common to veins and

country rock. He later analysed the small-scale folds produced by regional deformation in the laminated veins (Nicholson, 1978), suggesting that laminae formed in repeated acts of precipitation, separated by intervals of shear. He showed that interlaminar slip in early stages of folding was accommodated through the crystal–plastic behaviour of the single-crystal, calcite plates of which laminae are made. Work hardening, however, evidently raised the resistance of the plates, so that later slip was facilitated by pressure solution instead. Both this pressure solution, and the slip accompanying it, were concentrated where the surfaces between calcite plates lay at high angles to the principal shortening strain. Consequently, the through-going stylolites that gradually evolved were sited in fold limbs, eventually extending across the vein to isolate one hinge from another.

Plates and laminae show much greater continuity in cross-sections of microfolds than in sections parallel to fold axes. This condition, appears to be primarily related not to folding, however, but to the earlier-formed lineation, which in this vein lies approximately at right-angles to fold axes. In effect, calcite plates are elongated parallel to the lineation, which is so well developed on interlaminar surfaces through these veins.

Calcite laminae are separated from one another by thin seams composed of muscovite and fine-grained quartz. The grain size of the latter may be a product of recrystallization, rather than being primary. It may be significant that this material coats the grooves and ridges cut into the calcite plates. As far as is known, however, this recrystallization may have occurred during folding, rather than at the time of formation of the lineation. Muscovite flakes, for the most part, lie parallel to the seams.

Interpretation

The interest at this site relates principally to the laminated calcite veins. The platy and lineated nature of these is itself unusual, and the veins record deformation which pre-dates the main Caledonian phase.

Both the development of the primary fabrics of the laminated veins, and the nature of the folds later developing in them, are phenomena of interest in their own right. This interest is heightened by the apparent rarity of platy, calcite fabrics like those here, even in laminated veins. Investigation of the primary character of the veins also provides an opportunity to investigate structural development at times before the regional folds and cleavage had formed. The swing of the cleavage and folds to the ESE trend at this site is of considerable regional interest.

Veins composed of primary, platy, calcite crystals, and their distinctive folds, in Britain appear to be restricted to rocks of Wenlock or Ludlow age. Nicholson (1966) has pointed out the existence of veins with platy calcite in Ribblesdale and the southern Lake District, in country rocks of similar age and sedimentary facies to those of the Llangollen Syncline. Warren *et al.* (1970) have made similar observations on the Silurian rocks of northern Clwyd. The veins of western Caledonian Wales (Fitches *et al.*, 1986), in older host rocks, although laminated and lineated like those of the Llangollen Syncline are not composed of calcite but ferroan dolomite, in which the platy morphology is unknown.

Laminated and lineated veins also apparently made up of calcite plates have been reported, however, from slates of the Appalachian fold belt of Pennsylvania (Beutner *et al.*, 1977). This occurrence resembles in several ways that of Ca'er-hafod. The country rocks, for instance, are of similar mid- to late-Silurian age. The principal veins are similarly parallel to bedding and strongly folded. At the same time, fold hinges are separated from one another by stylolites cutting across veins, as at Ca'er-hafod. There is also a set of associated, discordant and boudined calcite–quartz veins. Although vein carbonate is described as calcite, nothing is said of its morphology. But the appearance of the folded veins, in the only figure showing the folded fabric in any detail, is quite compatible with a platy form for the calcites. Two points made by Beutner *et al.* (1977) are of special interest. Firstly, it is said that the Pennsylvanian bed-parallel veins lie along faults with only small displacement; however, no evidence is given. Secondly, the Pennsylvanian laminated veins apparently occur only in the gently dipping limbs of overturned folds. This is a contrast with Ca'er-hafod, where bedding–cleavage relationships suggest that veins are exposed in the region of a fold hinge.

Accepting the evidence of the discordant veins at Ca'er-hafod, the platy shape of the calcites may be described as a habit. The plates do not seem to be bounded, that is, by the compromise surfaces developed between adjacent crystals in competitive growth (Grigor'ev, 1965; Dickson, 1983). Such a habit has been described from crystal cavities in the New Jersey zeolite region of the USA (Schaller, 1932). The habit has been described as being indicative of high-temperature growth. This is of interest as the veins of the Llangollen Syncline are accompanied by well-crystallized muscovite (see Hamdi Lemnouar, 1988 for analysis). This proposal does not fit, however, with a source of mineralizing fluids in water drawn from the sediment body itself (Fitches *et al.*, 1986), at a time before even low-temperature metamorphism had begun.

Questions are raised by the lamination of the veins, but detailed explanations have yet to appear. However, using published analyses, some proposals may be outlined. The laminated veins, for instance, may be complex examples of the crack–seal veins of Ramsay (1980); that is, veins formed by successive development of microcracks followed by successive mineral infilling. The modified version of this hypothesis proposed by Cox (1987), seems to be particularly apt, designed as it was to explain veins forming when large displacements were occurring parallel to vein margins.

The formation of the lineation offers particular difficulties. The lineation is cut deep into the surfaces of only very thin calcite plates. If the incision of the lineation were mechanical, it is difficult to understand how the mechanically anisotropic and weak calcite plates were not at the same time deformed plastically and even broken into pieces. However, there is no sign of such disruption. The process, therefore, seems more likely to have occurred through sculpting by diffusion-based processes, rather than abrasion. The lineation, in effect, may have been formed through pressure solution, the surfaces representing an unusual variety of stylolite (Ramsay and Huber, 1987, p. 655).

The site lies at the eastern end of the fold cleavage arc that characterizes the Caledonian tectonic trend in North Wales (Shackleton, 1969). As Figure 4.1 shows, the trend of folds and cleavage at this end of the arc is slightly south of east; in Snowdonia (for example, Alexandra Quarry to Capel Curig, above) it swings to the NE–SW 'Caledonoid' trend, whereas to the south (Tan y Grisiau) it becomes almost N–S and then returns through NNE–SSW (Rheidol and Ponterwyd) to NE–SW (at Traeth Penbryn). This swing has been attributed by Shackleton (1969) to moulding against basement fault blocks, by Helm *et al.* (1963) to late (post-main-phase) Caledonian deformation, and most recently by Soper *et al.* (1987) to, once again, control by the basement during the main-Caledonian (early-Devonian) phase of closure of Iapetus.

Conclusions

This site is important from two points of view. Firstly, it provides an example of the Caledonian structural trend in north-eastern Wales, where it swings to a trend slightly south of east. This trend, which contrasts with other areas in Wales, has been explained as being a result of how the Palaeozoic rocks (at this locality of Silurian age) were compressed above and against the rigid basement of older (Precambrian) rocks.

The laminated and lineated calcite veins, that are extensively developed parallel to bedding, indicate an early stage of tectonic movement in the late Silurian. Their growth requires vertical opening along bedding, but other features suggest horizontal displacements. Although the growth of these unusual platy calcite veins is not fully understood, currently they are the subject of considerable interest. It is quite clear, however, that these veins were deformed in the (post-Ludlow) main phase Caledonian deformation of the Welsh Basin, when regional folding and cleavage were developed.

LLIGWY BAY (SH 49308746–49408803)
R. Scott

Highlights

Lligwy Bay contains a rare example of (presumed) Devonian rocks deformed during the Caledonian Orogeny. As this is the only Devonian locality in North Wales, these rocks provide unique information with which to assess the duration of the orogeny in this region.

Introduction

The Devonian rocks in Lligwy Bay record polyphase deformation, involving folding, cleavage formation and thrusting. The two upper Old Red Sandstone Group formations, that is the Porth-y-mor Formation and the Traeth Lligwy Formation (Allen, 1965), lie in a broad synclinal structure. This open structure has a monoclinal fold on its northern limb; on the southern limb thrusting and tight, minor folds with axial-planar cleavage occur (Figure 4.19).

The locality was described by Greenly (1919) in his memoir of Anglesey. A sedimentological interpretation of the Old Red Sandstone by Allen (1965) included some brief comments on the structure. Bates (1974) made a reconnaissance survey of the site and confirmed the observations of Greenly (1919). A number of large-scale tectonic interpretations have used the available information on Lligwy Bay, including those of Nutt and Smith (1981) and Woodcock (1984a), and the site has also appeared in a field guide (Bates and Davies, 1981), but no detailed, modern structural interpretation has been published. No fossils have been recorded from the sequence and its Devonian assignment is based upon lithological and stratigraphical similarities with other localities in south and south-east Wales.

Description

The site consists of a varied sequence of Devonian sediments exposed in a series of low cliffs on the north side of Lligwy Bay as far north as Trwyn Porth-y-mor. The strike of bedding and cleavage is approximately E–W. The site is described from north to south and the structure is depicted on a cross-section (Figure 4.19).

On the small headland opposite Trwyn Porth-y-mor, beds dip at <25° toward the south, and the dominant cleavage (S_1) dips between 50–70° to the north (Figure 4.20). In common with the site as a whole, cleavage is better developed in finer-grained siltstones and calcrete layers, where cleavage surfaces may be spaced closer than 5 mm. In sandstones, the spacing may be up to 0.20 m. In conglomerate units the cleavage is not clearly discernible. A second, localized cleavage spaced at 0.03–0.05 m offsets S_1 surfaces and dips at shallow angles to the north.

Towards the south, bedding becomes steeper, dipping to the south. It is locally overturned (around SH 49408787). Concurrent with this steepening of bedding, the principal cleavage becomes less steep, dipping at <30° to the north. To the south of this location, bedding returns to a shallow southerly dip, thus defining the monoclinal structure identified by Greenly (1919), and cleavage to a more steeply north-dipping attitude.

Bedding flattens out progressively to the south of the monocline, so that beds are undulating around horizontal (at SH 49428774). A low-angle surface exposed in the wave-cut platform at this locality probably represents a small thrust. In addition, the undulating beds are affected by minor normal faults (displacements <0.5 m) which post-date the cleavage and have a variety of attitudes. Immediately to the west, (SH 49398774),

S   N

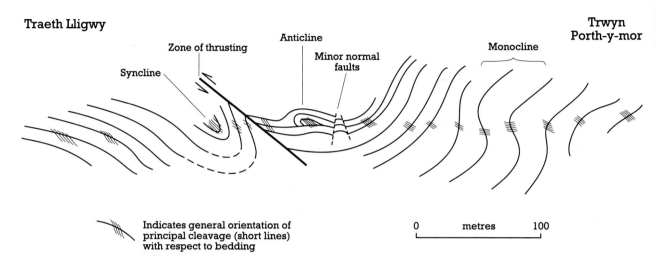

Figure 4.19   Sketch section illustrating the structure of the Devonian rocks on the north side of Lligwy Bay.

a small anticlinal hinge is exposed to which the principal cleavage is axial planar (dipping 54°N). The axis of the fold plunges gently east. This structure is the 'sharply over-driven anticline' featured in Greenly (1919; Figure 282, p. 586). The lower limb of the anticline is faulted out and, to the east, the anticline is replaced by a low-angle discordance of bedding, similar to the thrust surfaces observed close by. The fault surface has a parallel fabric which is similar in appearance, but oblique to the axial planar cleavage in the fold core; the contact between the two fabrics is not a distinct break.

From this point southwards, beds dip consistently north, first at shallow angles and then more steeply. A zone of thrusting at least 3 m wide and dipping at 40° to the north (SH 49338768) separates shallow north-dipping beds to the north from more steeply north-dipping beds to the south (Greenly, 1919; Figure 284, p. 586). In the footwall to the thrust, a tight synformal fold core can be observed (at SH 49328766). Again the principal cleavage is axial planar to the fold, dipping at ~50°N. The fold axis plunges gently east. To the south of this location, bedding dips at variable angles to the north.

Interpretation

On Anglesey, deformation of the Old Red Sandstone (which is presumed to be of late Silurian to early Devonian age) occurred before the deposition of the overlying Carboniferous Limestone succession (Allen, 1965). This, in conjunction with the site's location to the north of the Hercynian front, indicates that the deformation is not Hercynian (unless it is a freak local deformation) and therefore Caledonian, and that this phase was a post-Old Red Sandstone one. The deformed, presumed Devonian, succession at Lligwy Bay therefore provides crucial information for estimating the duration of the Caledonian Orogeny in North Wales.

A recent structural interpretation of the site is not available. However, its importance arises from the presence of deformation, rather than from its detailed interpretation. A number of important age relationships were established at Lligwy Bay by Greenly (1919), to which several additions can be made. These relationships indicate the polyphase nature of deformation. Greenly observed that cleavage was axial planar to the tight folds, but that it changed its orientation around the monocline, indicating that the monocline was a later structure. He also concluded that the thrusting post-dated the isoclinal folding because thrust surfaces are at a lower angle than, and they truncate, the cleavage. Although they may not be entirely synchronous, the spatial association between the tight folds and the thrusts does suggest that they are both a manifestation of the same deformation event. However, Greenly preferred to relate thrusting to monocline development which post-dates the

Figure 4.20 Lligwy Bay, Anglesey. Strongly developed, spaced cleavage in ?Devonian siltstones dips to the north in the hinge of a south-facing monocline. (Photo: J. Treagus.)

main cleavage (and may be related to the sporadic second cleavage). Two observations which Greenly did not make are that the principal cleavage maintains a northward dip across the broad synclinal structure and that relationships between bedding and cleavage change on the northern limb. These two observations suggest that the syncline may partly pre-date the cleavage and is therefore the first recognizable structure.

Deformation of the Old Red Sandstone at Lligwy Bay is, therefore, locally intense and polyphase, with possibly three phases of deformation and the development of thrusts. The deformation is, however, localized and, as Bates (1974) points out, is completely absent from the southern part of the inland outcrop. This variable intensity was interpreted by Bates as a reflection of basement control, deformation being most intense in the vicinity of major pre-existing faults.

The general consensus of opinion has been that the principal folds and cleavage in North Wales were produced in approximately end-Silurian times (for instance, Dewey, 1969; Coward and Siddans, 1979). A pre-Devonian age for the main

deformation was based on the observation that the Devonian lies unconformably on older rocks in the Welsh Basin and elsewhere. On Anglesey, the Devonian lies unconformably on the Ordovician, the unconformity post-dating the Bodafon Thrust. This indicated to Bates (1974) that the Devonian was deposited after the main period of deformation, which was end-Silurian. However, other authors have maintained that the main deformation period extended into the early Devonian (for example, Shackleton, 1953; Jones, 1955; Woodcock, 1984a). The significance of the unconformity at the base of the Devonian is open to doubt, as it may be the result of pre-Silurian, rather than just pre-Devonian erosion (George, 1963). The evidence at Lligwy Bay is limited by the lack of a firm age for these rocks, but it supports the conclusion that the lower part of the Old Red Sandstone suffered Caledonian movements. These movements were at least as intense as those affecting the adjacent Ordovician, and they have the same sense of overturning, towards the south-east. Woodcock (1984a), after assessing the information available at Lligwy Bay, concluded that any deformation climax in late Silurian Wales probably extended at least through Pridoli time and into the early Devonian. Soper *et al.* (1987) and McKerrow (1988), assessing evidence from Wales and the Lake District, considered that the main end-Caledonian movements were probably Emsian in age, equivalent to the Acadian of the Canadian Appalachians.

Conclusions

Lligwy Bay contains sedimentary rocks, thought to have been deposited during the Devonian Period (approximately 410–360 million years before the present), which have suffered folding, thrusting (low-angle faulting) and cleavage (closely spaced, parallel fractures) all during the Caledonian mountain-building episode. The intensity of this deformation was at least as strong as that suffered by the Ordovician succession which underlies the Devonian rocks. The fact that Old Red Sandstone sedimentary rocks are deformed indicates that Caledonian deformation of some significance extended beyond the end of the Silurian and well into Devonian times.

CARMEL HEAD
(SH 29079279–30709300)
D. E. B. Bates

Highlights

This is the only well-exposed and clearly identifiable, major low-angle thrust in the Welsh Caledonides. The Carmel Head Thrust, overriding an earlier steep fault and cut by later faults, can be traced eastwards for some 400 m in the cliffs and well-exposed ground to the south; it thrusts Precambrian schists southwards over Caradoc Series shales. Further east, the Ordovician Garn Breccia and overlying shales are overridden by Precambrian rocks on the Mynachdy Thrust.

Introduction

The Carmel Head region is well known for both its stratigraphical and, in particular, its structural interest. The fault relationships were first described clearly by Callaway (1884), and his conclusions were confirmed and amplified by Matley (1901) and Greenly (1919, 1920), who mapped the area

on the scale of 1:2500. Greenly (1919, pp. 547–9) considered that the Precambrian Mona Complex (Monian) had been overthrust on to the Ordovician along the E–W-striking Carmel Head Thrust by about 20 km. Bates (1972, 1974) showed that the palaeogeographical arguments upon which this figure was based were incorrect. Other associated thrusts include the Mynachdy Thrust and thrusts within the Ordovician shales at this site, described by Greenly (1919), Bates (1972, 1974), Bates and Davies (1981) and Barber and Max (1979).

Description

The area shows a sequence of complexly folded and faulted units. South of Porth Ogo'r geifr there are gneisses, thought by Greenly (1919) and Barber and Max (1979) to be part of an Archaean basement, and by Shackleton (1969) to be high-grade members of the late Precambrian (Cadomian) Monian. These are faulted against cleaved Caradoc Series shales. At a small cove immediately west of Porth y Wig, the Caradoc rocks are overthrust by the Monian Amlwch Beds of the New Harbour 'Series': this is the type locality for the Carmel Head Thrust (Figure 4.21).

On Carmel Head itself, Gwna Mélange of the Mona Complex (Monian) to the east is unconformably overlain by the rather similar Ordovician Garn Formation. Towards Porth Newydd, the Garn breccias pass up gradually into graptolitic shales, with one thrust slice of Monian phyllites in the sequence. Finally, at Porth Newydd, the Church Bay Tuffs of the Monian are thrust over these shales along the Mynachdy Thrust. The Church Bay Tuffs are probably intermediate in age between the Amlwch Beds and the younger Gwna Mélange in the Monian. The principal features of the site (Figure 4.22) are described from west to east.

Porth Ogo'r-geifr

The inlet is eroded in an E–W-striking fault complex with a steep, northerly dip. On the south side (SH 29149279) sandstones and shales of the Garn Formation are exposed, faulted against the Gader Gneisses to the south. The fault is mineralized: quartz-veined gneiss and shales are exposed in a small excavation around an old shaft opening 40 m east of the head of the inlet.

Figure 4.21 Carmel Head, Anglesey. Figure standing on the low-angle fault plane which has thrust Precambrian schists over Ordovician shales. (Photo: J. Treagus.)

Porth Ogo'r-geifr to the Thrust Inlet (SH 29579303, just west of Porth y Wig)

Well-exposed, cleaved Caradoc shales make up the cliffs, which gradually decrease in height to the north-east. A well-developed crenulation cleavage (first described by Greenly in 1919) is present, dipping at low angles to the south. This is spatially related to the thrusts, and was described by Bates (1974) as being linked to the thrust movements. It is only rarely possible to determine bedding, where lithological variations are found, for instance, in the north wall of Porth y Dyfn (SH 29399287). Here, there is a debris-flow breccia of extremely angular blocks of schist and phyllite (up to 0.60 m long) in a shale matrix. A low-angle north-dipping thrust within the shales is well exposed on the coastal rock platform (SH 29449298). There tension gashes and slickencrysts both confirm the southerly movement of the hanging wall, and the crenulation cleavage is well developed.

The Thrust Inlet

The inlet is excavated along a NNW–SSE-trending high-angle fault, which displaces the Carmel Head Thrust at the head of the inlet. The thrust itself follows the north wall, where the chloritic schists of the Precambrian Amlwch Beds overlie Ordovician shales, with the fault contact dipping gently north. The thrust cuts an earlier high-angle fault towards the low water mark. It is possible that both high-angle faults are part of the thrust sequence, although Bates (1974) interpreted the earlier one as being part of a pre-thrust phase of high-angle reverse faulting.

Porth y Wig and Carmel Head

The Amlwch Beds of the New Harbour 'Series' form the hanging wall of the thrust, and the south side of Porth y Wig, but the headland (Garn Mynachdy) of Carmel Head itself is formed of

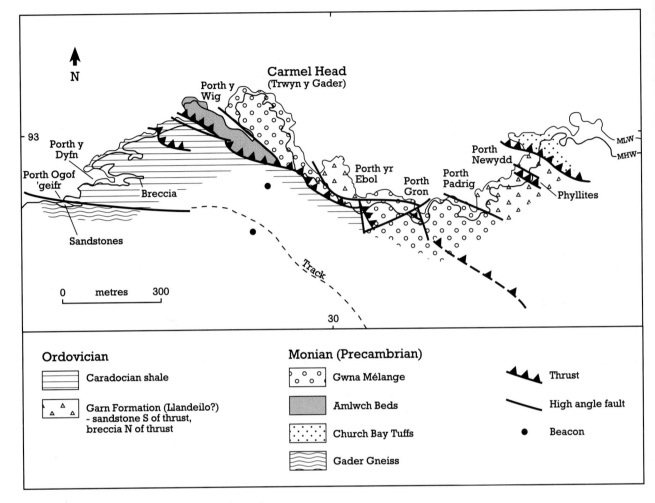

Figure 4.22 Geology of the Carmel Head site.

Gwna Mélange; its contact with the Amlwch Beds is another fault. The mélange is predominantly siliceous, but a band of carbonate mélange is also present, with clasts of dolomite and limestone up to 2–3 m long – which may represent the remains of a dismembered carbonate horizon in the parent succession.

Porth yr Ebol–Porth Gron (SH 29909288)

Three rock units are present in this section, from south to north:

1. Caradoc shales beneath the thrust.
2. Gwna Mélange on the hanging wall.
3. Llandeilo (?) Garn Formation breccias, either faulted against the Gwna Mélange, or unconformable on it.

The main thrust appears in at least five locations:

1. SH 29909288 as a small faulted exposure in the cliff.
2. SH 29969283 in a natural cave.
3. SH 30009280 and east of it along the wave-cut platform.
4. In the east wall of Porth yr Ebol.
5. On the wave-cut platform in Porth Gron.

Again the thrust is cut by ENE–WSW and N–S-trending faults. In Porth yr Ebol the Garn Formation is faulted against the Gwna Mélange, but just east of this inlet (SH 30129278), it appears to be unconformable on the mélange, although as the two formations are similar in character, the boundary is difficult to trace.

Porth Padrig to Porth Newydd

In Porth Padrig, another WNW–ESE fault, dipping steeply to the north (the Padrig Slide of Greenly, 1919), separates the Gwna Mélange from the Garn Formation to the north. The Garn Formation fines upwards (Bates, 1972) into shales with thin breccia and sandstone beds. A mass of phyllites is thrust up into the shales (SH 30649287). In Porth Newydd the Caradoc shales are complexly faulted on a small scale, and are overthrust on the Mynachdy Thrust by Church Bay Tuffs of the Monian.

Interpretation

The sequence described here is of both historical and current interest. It forms the type area or exposure of the Carmel Head Thrust, and there is here no doubt of the reality of this structure. Bates (1974), however, points out that Greenly's (1919, pp. 541-557) arguments for a 20 km southward translation on the thrust, were based on a palaeogeographical interpretation of the Ordovician, which is no longer tenable (Bates, 1972). Although Bates (1974) confirmed the N–S direction of transport from growth fibres and the orientation of the crenulation structure related to the movements, he had already observed (1972, p. 55) that the minimum movement required on the thrust is only in the order of metres.

At Carmel Head there are two low-angle faults which clearly emplace older (Monian) rocks over younger (Ordovician). There are also other parallel, low-angle faults that are probably thrusts, as well as high-angle reverse faults. Elsewhere on Anglesey the plane mapped as the equivalent of the Carmel Head Thrust is usually a high angle reverse fault. It is probable that remapping of this area, in conjunction with modern ideas on thrust geometry, may lead to revision of some of the faulting sequence, but should make the zone of even greater significance.

Models for the evolution of the structure of the Welsh Basin (for example, Shackleton, 1969; Coward and Siddans, 1979) appeal to thrusting, related to mid-crustal décollement, as a major response to the main-phase Caledonian shortening (late Silurian to early Devonian). The movements described here, although possibly closely following, certainly post-date the main-phase cleavage and folding. The Old Red Sandstone rocks at Lligwy Bay (see above) are also affected by

southward thrusting which also appears to post-date, but to be closely associated with cleavage and folding of the same main phase. This thrusting on Anglesey appears to pre-date the Carboniferous succession.

Descriptions of similar thrusts elsewhere in the north Welsh Basin are rare. The most famous, the Tremadoc Thrust (Fearnsides, 1910), has been shown by Smith (1987, 1988) to be a pre-lithification structure. The thrusting at Trum y Ddysgl (see above) has the same south-easterly sense of movement but is directly related to the folding and cleavage.

Conclusions

Carmel Head provides one of the clearest, and the only convincing, examples of late thrusting in the Caledonides of Wales. Thrusts (faults lying at a low angle to the horizontal) were a major product of crustal shortening caused by compression in the Caledonian event. The amount of movement on this particular major dislocation has been the subject of dispute, but it thrusts already cleaved and folded Precambrian basement rocks south over Early Palaeozoic rocks. The thrust movements post-date the folding and cleavage which were formed by the main tectonic events of the Caledonian Orogeny, and this has significance in establishing a chronology for this period of mountain building. The clarity of the exposures hereabouts makes the region of paramount importance.

LLANELWEDD QUARRY (SO 051522)
N. H. Woodcock

Highlights

Llanelwedd Quarry provides the best-exposed section across the structures of the Pontesford Lineament, one of the major fault belts that was active between the Welsh Basin and the Midland Platform of England in Palaeozoic times.

Introduction

This working quarry exposes rocks of the Builth Igneous Complex (Llanvirn Series) of early Ordovician age, at the southern end of the Builth–Llandrindod Inlier. This inlier exposes a 'window' of Llanvirn and Llandeilo sedimentary and volcanic rocks, unconformably overlain by Upper Llandovery and Wenlock strata. Historically, most interest has focused on the volcano-sedimentary stratigraphy of the locality (Elles, 1940; Jones and Pugh, 1941, 1946, 1949; Furnes, 1978). In a structural context, this area is more important for the numerous fault zones that cut the section. These were first mapped by Jones and Pugh (1949) and interpreted as part of a strike-slip fault system by Jones (1954) and Baker (1971). The site lies in the zone of structures (Woodcock, 1984b), known as the Pontesford Lineament, which extends from the Cheshire Basin to South Wales (Figure 4.1). Although the lineament was intermittently active at least from mid-Ordovician times through to the Triassic, the principal fault displacements, both strike-slip and dip-slip, were late Ordovician to early Silurian (pre-Upper Llandovery unconformity). The best-documented displacements of up to 5 km, dextral movements, occur on faults in the Shelve Inlier, some 40 km north-east of the present site (Lynas, 1988). Woodcock (1984b) estimates that the total (dextral) movement in the zone could be in excess of 20 km. The locality has been used (Woodcock, 1987b) as an example of the structural architecture of a strike-slip fault belt.

Description

The geology of the quarry complex is summarized in Figure 4.23, based mainly on exposures in four main arcuate working faces, each 20 m high. The map shows the position of the faces in 1984, but they are being continually worked back northwards. To remove the confusing effects of the stepped topography, the map is constructed as a projection on to a hypothetical smooth surface through the top of each face.

The lithological sequence is dominated by basalt lavas, mostly feldspar-phyric and highly vesicular. These contain intercalated agglomerates, a sandstone body and a dolerite sill. This structurally conformable sequence dips moderately westwards. It is cut by numerous faults, mostly striking approximately north, and dipping steeply eastwards. Three major zones of faults can be recognized, summarized as an inset on Figure 4.23. A central zone mainly comprises NNE–SSW striking dip-slip faults. This zone separates eastern and western zones containing mostly strike-slip faults with the same strike. Minor E–W-striking strike-slip faults also occur. These faults dominate

the overall structure, and about 57% of them are strike-slip, 31% oblique-slip, and only 12% are dip-slip faults. Fault slip directions can be determined from cataclastic slickensides and from slickenfibres, elongate crystal growths in the fault planes. The sense of slip can rarely be determined directly from these structures. Offsets of distinctive stratigraphical units can be used to obtain dip-slip senses, but are unreliable for most strike-slip faults, because the strike directions of bedding and faults are so close. The limited data show that most of the dip-slip faults have normal rather than reverse offsets, but that sinistral and dextral offsets on strike-slip faults are equally numerous. Many of the west-dipping 'bedding' surfaces, mainly boundaries of lava-flow units, have also acted as displacement planes. They show northerly strike-slip slickenlines.

Interpretation

The faults in Llanelwedd Quarry record an important strike-slip faulting event along the Pontesford Lineament. This event cannot be dated at this locality, but mapping of the whole Builth–Llandrindod Inlier (compilation by IGS, 1977) shows that many of the faults of the strike-slip system do not cut the unconformable Upper Llandovery (Lower Silurian) and younger cover. Regional evidence suggests an Ashgill (Late Ordovician) age as most likely (Woodcock, 1984b). Llanvirn and Llandeilo rocks in the Builth Inlier are displaced by the faults and Caradoc sequences do not match across the main fault in the Pontesford area. Later reactivation of the Pontesford Lineament is suggested north-east of the Builth area by its coincidence with the fold–fault zone of the Clun Forest Disturbance, affecting rocks as late as those of the Pridoli.

Although the kinematic interpretation of the Llanelwedd faults is made uncertain by the paucity of data on slip sense, Woodcock (1984b) has proposed that the main fault sets (shown in Figure 4.23) interact to form a linked system capable of accommodating three-dimensional bulk strain. Woodcock (1987b) suggests that the NNW–SSE strike-slip faults are sinistral and that they have

Figure 4.23 Geological map of the main Llanelwedd Quarry with inset summary of main kinematic zones (after Woodcock, 1987b).

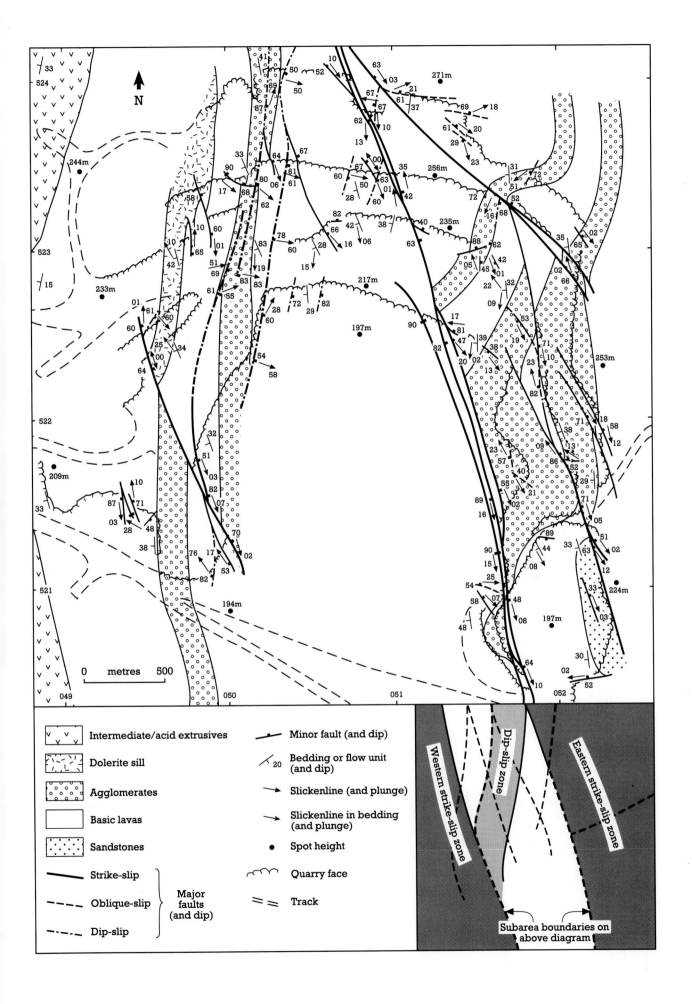

played an antithetic role to more major, dextral, north-east-striking faults mapped beyond the quarry exposures (Jones and Pugh, 1949). These dextral faults seem to splay off the major Cwm Mawr Fault that forms the main element of the Pontesford Lineament within the inlier. Dextral faulting of late Ordovician age has also been suggested further along the lineament in the Shelve Inlier (Woodcock, 1984b; Lynas, 1988).

The Ashgill deformation event may have been responsible for juxtaposing the Welsh Basin against the Midland Platform from former, more distant positions (Woodcock and Gibbons, 1988). The suggestion of a dextral sense contrasts with the mainly sinistral displacements which are deduced from evidence across Wales of the main Acadian (late-Caledonian) deformation in late Silurian to early Devonian times. Due to generally poor, natural exposure along the Pontesford Lineament there are very few localities where the evidence for strike-slip is well displayed. Because it is an actively working quarry, Llanelwedd is presently the best-exposed locality. It is likely to remain an important site for testing changing hypotheses on the nature of the lineament.

The good constraints on the three-dimensional geometry of the structure at Llanelwedd give insights into the working of strike-slip fault systems in general. Of particular interest is the way in which four main fault sets (shown in Figure 4.23), including the bedding-parallel slip, interact to form a linked system capable of accommodating three-dimensional bulk strain (Woodcock, 1987b). The steep, NNW strike-slip faults dominate, with significant strike-slip on the westerly dipping bedding surfaces and bedding parallel faults. A zone of steep northerly striking dip-slip faults links two of the strike-slip strands and there is a weaker easterly striking set of strike-slip faults. When these four sets of faults are rotated so that the regional bedding is horizontal, three become vertical and one (parallel to bedding) horizontal, presumably their original attitude in late-Ordovician to early-Silurian times. The faults can then be seen as part of a linked, dextral, strike-slip system with accommodation of strain on to smaller dip-slip and bedding parallel faults. The locality is excellent for further detailed investigation of the mechanics of this sort of fault system.

Conclusions

Llanelwedd Quarry is important as a well-exposed locality through the major fault belt known as the Pontesford Lineament. This fault belt was an active zone of dislocation between the Midland Platform and the Welsh Basin during Palaeozoic times. The rocks seen in the quarry are igneous (volcanic and intrusive) rocks of early Ordovician age. The site demonstrates the importance of strike-slip faulting during a deformation event, probably during the Late Ordovician Period (Ashgill), that might have involved large lateral displacements along the line of the lineament. The geometry of the fault system at the locality is also of some general interest in understanding the mechanics of strike-slip fault movements. This is an important site that allows observations on an otherwise poorly exposed feature, which is one of Britain's major Caledonian tectonic structures.

DOLYHIR QUARRIES, OLD RADNOR (SO 245581)
N. H. Woodcock

Highlights

These quarries provide a section through the Church Stretton Fault Zone, one of the active structures between the Welsh Basin and the Midland Platform of England during Palaeozoic time. They expose the clear angular unconformity of Wenlock Series strata on the Precambrian, important in providing dates which constrain interpretation of the regional tectonic history.

Introduction

The complex of quarries around Dolyhir lies within the Old Radnor Inlier, a small fault-bounded and fault-dissected sliver along the south-west continuation of the Church Stretton Fault. The inlier comprises Precambrian sedimentary rocks unconformably overlain by Lower Wenlock limestones. Wenlock age shales surround the inlier, mainly with faulted contacts but, in places, possibly with depositional contacts. The main structural features of the locality are numerous steep faults, mostly cutting both Precambrian and Wenlock rocks.

Early interest in the inlier (Callaway, 1900)

focused on the presumed Precambrian sediments and their possible correlation with the Longmyndian of Shropshire. This correlation was supported by later work (Garwood and Goodyear, 1918) which also showed the palaeontological interest of the overlying limestones, and their correlation with the Woolhope Limestone (Lower Wenlock) further east. The recognition of important faulting within the inlier and more detailed work (Kirk, 1951, 1952) produced the model that the inlier was a basement block upthrust along the Church Stretton Fault Belt in Caledonian or later times. The inliers along this fault were important evidence for the theory that the main, NE–SW, 'Caledonoid' lineaments in South Wales and the Borderland overlie steep fault belts that cut the basement to some depth (Owen, 1974; Owen and Weaver, 1983; Woodcock, 1984a). More recent work (Woodcock, 1988) has shown that the post-Wenlock faults in the inlier have dominantly strike-slip rather than dip-slip displacements.

Description

The main geological features of the complex of old and working quarries at Dolyhir in 1985 are summarized in Figures 4.24 and 4.25. Strinds Quarry is currently being worked south-westward within the limestone only. Dolyhir Quarry is being extended eastward at all levels. In Strinds Quarry (Figure 4.24), Precambrian sandstones and conglomerates of the Strinds Formation are exposed in the lower faces, dipping steeply to the north-west. Gently dipping Wenlock Dolyhir Limestone unconformably overlies the Strinds Formation, with a patchily developed basal rudite. The three most continuous faults in the quarry strike NNE–SSW and dip steeply to the WNW. They cut both Precambrian and Wenlock and displace the unconformity surface with normal offset (downthrow to the WNW). However, slickensides on these faults all indicate strike-slip displacements. The displacement sense is mostly indeterminate. Common, minor strike-slip faults parallel the continuous faults.

Minor WNW–ESE or NW–SE striking faults are common above the unconformity, and they are dextral where the sense can be determined from stepped slickenfibres. This fault set is interpreted as conjugate to the main northerly set, forming the Riedel shear pattern common in strike-slip systems (see Figure 4.24 inset). On this basis, the main

faults would have a sinistral sense. Minor dip-slip faults strike between NNE–SSW and NE–SW, and show an extensional component both above and below the unconformity.

In Dolyhir Quarry (Figure 4.25), Precambrian rocks again form the lower faces. Strinds Formation sandstones outcropping in the south-east are faulted against the finer-grained, better-bedded sediments of the Yat Wood Formation in the central and north-eastern parts. This fault zone strikes NE–SW and contains mostly south-east-dipping faults, some showing strike-slip and some dip-slip slickensides. Several subparallel faults cut the eastern part of the Yat Wood Formation, mostly showing dip-slip displacements. None of these NE–SW or E–W striking faults unambiguously cuts the Wenlock unconformity and therefore they could be pre-Wenlock, even Precambrian, in age. They all abut against, or anastomose with, a major, NNE–SSW-striking fault (SO 24425828 to 24475845) which displaces the unconformity down by about 30 m to the WNW. The central segment of this fault shows dip-slip slickensides. West of the fault, the Yat Wood Formation and Dolyhir Limestone are both cut by two subparallel major faults, one evidencing strike-slip, the other some additional dip-slip component. The unconformity surface, everywhere overlain by basal rudites, is progressively downfaulted towards the north-west part of the quarry. As in Strinds Quarry, the Dolyhir Limestone commonly shows steep NW–SE-striking strike-slip faults with dextral sense. Although the Riedel shear pattern is not so clear here it is still compatible with sinistral shear on the main northerly strike-slip faults (see Figure 4.25 inset).

Complementing the information from Strinds and Dolyhir Quarries are numerous minor exposures throughout the Old Radnor Inlier and also the extensive Gore Quarry at its north-west end. This is of less stratigraphical interest because it contains only Precambrian rocks, but it contains a suite of faults comparable with those described above (details given by Woodcock, 1988).

Interpretation

The geometrical pattern of structures in the Old Radnor Inlier is exceedingly complex and some aspects of its kinematic interpretation remain tentative. The clearest feature is the predominance of strike-slip faults over dip-slip. About 65% of faults affecting the Precambrian and over 80%

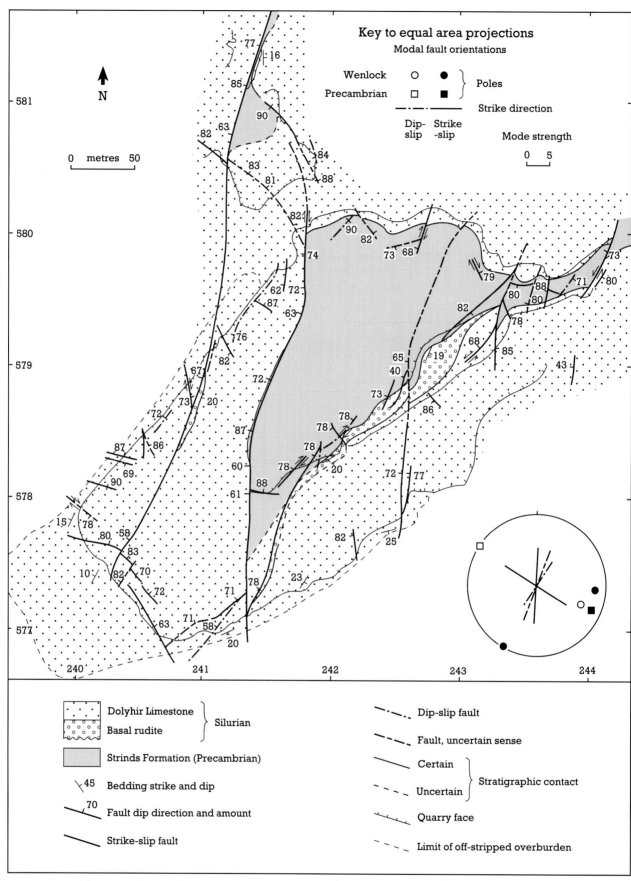

Figure 4.24 Structural map of Strinds Quarry with inset stereogram showing modal orientations of strike-slip and dip-slip faults (after Woodcock, 1988).

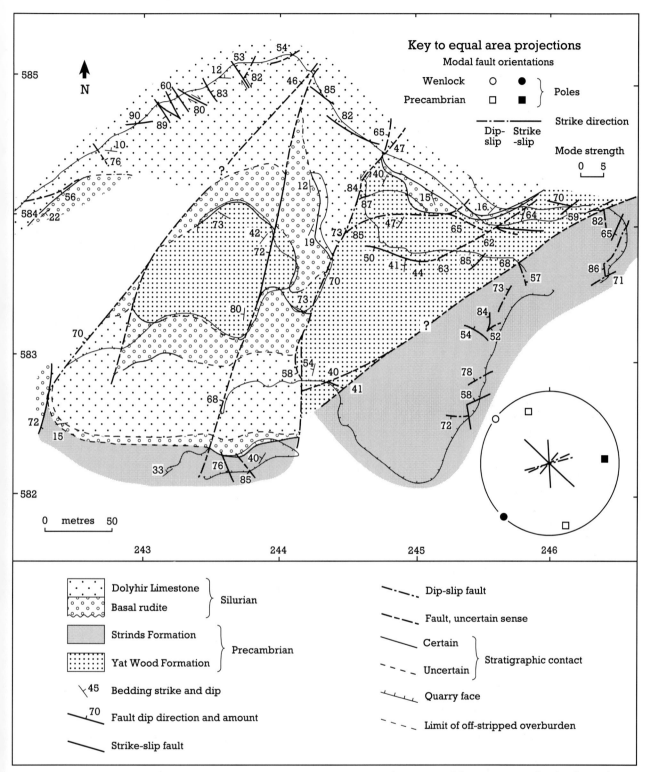

Figure 4.25 Structural map of Dolyhir Quarry with inset stereogram showing modal orientations of strike-slip and dip-slip faults (after Woodcock, 1988).

affecting the Wenlock show slickensides shallower than 45°. This result, taken with the steep attitude and braided interconnection of most faults, suggests that the inlier has suffered important post-Wenlock deformation in a regime with a strong transcurrent component. The lower proportion of strike-slip faults in the Precambrian is compatible with the post-Wenlock phase having reactivated earlier dominantly dip-slip faults.

Throughout the inlier, a main strike-slip fault set striking N–S, or NNE–SSW is accompanied by a subsidiary set striking NW or NNW. This pattern is most simply explained as a Riedel shear response to sinistral strike-slip deformation, parallel to the NE–SW trend of the inlier (that is, subsidiary faults are produced as a result of strain produced by the principal fault). Strike-slip displacement along the Church Stretton Fault is evidenced elsewhere along its length, principally in the Church Stretton area 40 km away, at its north-eastern end (Figure 4.1). Here (see Woodcock, 1988, for summary) some movements which can only be dated as pre-Llandovery and others which are post-Caradoc strongly suggest sinistral movements from their Riedel pattern, displacement of vertical beds and strike-slip slickensides. However, the Old Radnor Inlier provides the most direct and convincing evidence of sinistral movements which post-date the Wenlock.

The timing of the sinistral displacements is not well constrained. They could be late Silurian or early Devonian, and driven by the regional sinistral component to the main Acadian (late-Caledonian) deformation in Wales (Woodcock *et al.*, 1988). Alternatively, they could be Variscan or even later, since displacements of this age are evidenced along strike on the Church Stretton Fault. However, there is no direct evidence that the Church Stretton Fault ever accommodated very large lateral displacements (Woodcock and Gibbons, 1988). Where the offset can be measured in the Church Stretton area it ranges from a few hundred metres to 1.5 km (Greig *et al.*, 1968). Allowing for similar movements on faults of indeterminate offset, the total strike-slip across the whole zone probably lies in the range of 2–10 km (Woodcock, 1988). As ideas on the tectonics of southern Britain are further developed, the Dolyhir localities will remain an important source of relevant data.

Conclusions

The Dolyhir Quarries provide good sections through the Caledonian Church Stretton Fault Zone, in an area of otherwise poor exposure. The stratigraphical and structural relationships at the locality place constraints on the timing of displacements along this important tectonic boundary. It is a major tectonic lineament which has a long history of activity in the zone between the Welsh Basin and the English Midlands. These relationships demonstrate the involvement of old basement in the fault movements, and, in particular, show an important component of post-Wenlock strike-slip fault displacement. Although little can be said about the size of such displacements, there is considerable potential for future study of the part that this major lineament played in Caledonian earth movements.

References

Allen, J.R.L. (1965) The sedimentation and palaeo-geography of the Old Red Sandstone of Anglesey, North Wales. *Proceedings of the Yorkshire Geological Society*, **35**, 139–85.

Allen, P.M. and Jackson, A.A. (1985a) *Geology of the country around Harlech*. Memoir of the Geological Survey of the UK, HMSO, London, 112 pp.

Allen, P.M. and Jackson, A.A. (1985b) Geological excursions in the Harlech Dome. *Classical areas of British Geology*, British Geological Survey, HMSO, London, 94 pp.

Anderson, T.B. and Cameron, T.D.J. (1979) A structural profile of Caledonian deformation in Down. In *The Caledonides of the British Isles – reviewed*. (eds A.L. Harris, C.H. Holland, and B.E. Leake), Geological Society of London Special Publication, No. 8, pp. 263–7.

Anderson, T.B. and Oliver, G.J.H. (1986) The Orlock Bridge Fault: a major Late Caledonian sinistral fault in the Southern Uplands terrane, British Isles. *Transactions of the Royal Society of Edinburgh: Earth Sciences*, **77**, 203–22.

Anketell, J.M. (1987) On the geological succession and structure of south-central Wales. *Geological Journal*, **22**, 155–65.

Arthurton, R.S., Johnson, E.W., and Mundy, D.J.C. (1988) *Geology of the country around Settle*. Memoir of the Geological Survey of the UK, HMSO, London.

Aveline, W.T. (1872) On the continuity and breaks between the various divisions of the Silurian strata in the Lake District. *Geological Magazine*, **9**, 441–2.

Aveline, W.T., Hughes, T. McK., and Strahan, A. (1888) *The Geology of the country around Kendal, Sedburgh, Bowness and Tebay*. Memoir of the Geological Survey of the UK, HMSO, London.

Bailey, E.B. (1929) The Palaeozoic mountain systems of Europe and America. *Report of the British Association for the Advancement of Science*, Glasgow, 1928, Trans. Secs, 57–76 (British Isles, General).

Baker, J.W. (1971) Intra-Lower Palaeozoic faults in the southern Irish Sea area. *Geological Magazine*, **108**, 501–9.

Barber, A.J. and Max, M.D. (1979) A new look at the Mona Complex (Anglesey, North Wales). *Journal of the Geological Society of London*, **136**, 407–32.

Barber, A.J., Max, M.D., and Brück, P.M. (1981) Geologists' Association–Irish Geological Association; Field Meeting in Anglesey and south-eastern Ireland, 4–11 June, 1977. *Proceedings of the Geologists' Association*, **92**, 269–91.

Barker, A.J. and Gayer, R.A. (1985) Caledonide–Appalachian tectonic analysis and evolution of related oceans. In *The Tectonic Evolution of the Caledonide–Appalachian Orogen*. (ed. R.A. Gayer), F. Vieweg and Sohn, Braunschweig, pp. 126–65.

Barnes, R.P. (1989) *Geology of the Whithorn District*. Explanation of Sheet 2 (Scotland), Memoir of the British Geological Survey, HMSO, London, 20 pp.

Barnes, R.P., Anderson, T.B., and McCurry, J.A. (1987) Along-strike variation in the stratigraphical and structural profile of the Southern Uplands Central Belt in Galloway and Down. *Journal of the Geological Society of London*, **144**, 807–16.

Bassett, D.A. (1969) Some of the major structures of early Palaeozoic age in Wales and the Welsh

References

Borderland: an historical essay. In *The Pre-cambrian and Lower Palaeozoic rocks of Wales*.(ed. A. Wood), University of Wales Press, Cardiff, pp. 67–116.

Bates, D.E.B. (1968) The Lower Palaeozoic brachiopod and trilobite faunas of Anglesey. *Bulletin of the British Museum (Natural History): Geology*, **16**, p. 127.

Bates, D.E.B. (1972) The stratigraphy of the Ordovician rocks of Anglesey. *Geological Journal*, **8**, 29–58.

Bates, D.E.B. (1974) The structure of the Lower Palaeozoic rocks of Anglesey, with special reference to faulting. *Geological Journal*, **9**, 39–60.

Bates, D.E.B. and Davies, J.R. (1981) Anglesey. *Geologists' Association Guide*, **40**, 31 pp.

Bell, A.M. (1975) A finite strain study of accretionary lapilli tuff in the Borrowdale Volcanic Group, English Lake District. Unpublished Ph.D. thesis, University of Sheffield.

Bell, A.M. (1981) Strain factorisations from lapilli tuff, English Lake District. *Journal of the Geological Society of London*, **138**, 463–74.

Bell, A.M. (1985) Strain paths during slaty cleavage formation – the role of volume loss. *Journal of Structural Geology*, **7**, 563–8.

Beutner, E.C., Jancin, M.C., and Simon, R.W. (1977) Dewatering origin of cleavage in light of deformed calcite veins and clastic dikes in Martinsburg slate, Delaware Water Gap, New Jersey. *Geology*, **5**, 118–22.

Bluck, B.J. (1986) The Scottish paratectonic Caledonides. *Scottish Journal of Geology*, **21**, 437–64.

Boulter, C.A. and Soper, N.J. (1973) Structural relationships of the Shap Granite. *Proceedings of the Yorkshire Geological Society*, **39**, 365–9.

Bouma, A.H. (1962) *Sedimentology of Some Flysch Deposits: A Graphic Approach to Interpretation*. Elsevier, Amsterdam, 168 pp.

Bracegirdle, D.R. (1974) Structure and Stratigraphy of the Towyn–Abergynolwyn Area of Southern Merionethshire in Relation to the Bala Fault. Unpublished MSc thesis, University of Wales, Aberystwyth.

Branney, M.J. and Soper, N.J. (1988) Ordovician volcano-tectonics in the English Lake District. *Journal of the Geological Society of London*, **145**, 367–76.

British Geological Survey (1966) *Tectonic map of Great Britain and Northern Ireland*, 1st edn.

British Geological Survey (1979) *Geological Survey Ten Mile Map, South Sheet*, 3rd edn (Solid).

Bromley, A.V. (1963) The Geology of the Blaenau Ffestiniog Area, Merionethshire. Unpublished PhD thesis, University of Wales, Aberystwyth.

Bromley, A.V. (1969) Acid plutonic igneous activity in the Ordovician of North Wales. In *The Precambrian and Lower Palaeozoic rocks of Wales*. (ed. A. Wood), University of Wales Press, Cardiff, pp. 387–408.

Bromley, A.V. (1971) Phases of deformation in North Wales. *Geological Magazine*, **108**, 548–50.

Brown, P.E., Miller, J.A., and Soper, N.J. (1964) Age of the principal intrusions of the Lake District. *Proceedings of the Yorkshire Geological Society*, **34**, 331–42.

Callaway, C. (1884) The Archaean and Lower Palaeozoic rocks of Anglesey (with an appendix, on the petrology of the rocks by T.G. Bonney). *Quarterly Journal of the Geological Society of London*, **40**, 567–89.

Callaway, C. (1900) On Longmyndian Inliers at Old Radnor and Huntley (Gloucestershire). *Quarterly Journal of the Geological Society of London*, **56**, 511–20.

Campbell, S.D.G., Reedman, A.J., and Howells, M.F. (1985) Regional variations in cleavage and fold development in North Wales. *Geological Journal*, **20**, 43–52.

Campbell, S.D.G., Howells, M.F., Smith, M., and Reedman, A.J. (1988) A Caradoc failed-rift within the Ordovician marginal basin of Wales. *Geological Magazine*, **125**, 257–66.

Capewell, J.G. (1955) The post-Silurian pre-marine Carboniferous sedimentary rocks of the eastern side of the English Lake District. *Quarterly Journal of the Geological Society of London*, **111**, 23–46.

Cattermole, P. and Jones, A. (1970) The geology of the area around Mynydd Mawr, Nantlle, Caernarvonshire. *Geological Journal*, **7**, 111–28.

Cave, R. (1984) *Outline of geology*. From 1:50 000 geological map, Sheet 163 (Aberystwyth), British Geological Survey.

Cave, R. and Hains, B.A. (1986) *Geology of the country between Aberystwyth and Machynlleth*. Memoir of the Geological Survey of the UK, HMSO, London.

Clark, L. (1964) The Borrowdale Volcanic Series between Buttermere and Wasdale, Cumberland. *Proceedings of the Yorkshire Geological Society*, **34**, 343–56.

Clarkson, C.M., Craig, G.Y., and Walton, E.K. (1975) The Silurian rocks bordering Kirkcudbright Bay, South Scotland. *Transactions of the Royal Society of Edinburgh: Earth Sciences*, **69** 313–25.

References

Cobbold, P.R. and Quinquis, H. (1980) Development of sheath folds in shear regions. *Journal of Structural Geology*, **2**, 119–26.

Cook, D.R. and Weir, J.A. (1979) Structure of the Lower Palaeozoic rocks around Cairnsmore of Fleet, Galloway. *Scottish Journal of Geology*, **15**, 187–202.

Cooper, D.C., Lee, M.K., Fortey, N.J. *et al.* (1988) The Crummock Water Aureole: a zone of metasomatism and source of ore metals in the English Lake District. *Journal of the Geological Society of London*, **145**, 523–40.

Coward, M.P. and Siddans, A.W.B. (1979) The tectonic evolution of the Welsh Caledonides. In *The Caledonides of the British Isles – Reviewed.* (eds A.L. Harris, C.H. Holland, and B.E. Leake), Special Publication of the Geological Society of London, No. 8, pp. 187–98.

Cox, S.F. (1987) Antitaxial crack–seal vein microstructures and their relationship to displacement path. *Journal of Structural Geology*, **9**, 779–87.

Craig, G.Y. and Walton, E.K. (1959) Sequence and structure in the Silurian rocks of Kirkcudbrightshire. *Geological Magazine*, **96**, 209–20.

Craig, G.Y. and Walton, E.K. (1983) Asymmetrical folding in the Hawick Rocks of the Galloway area, Southern Uplands: comment. *Scottish Journal of Geology*, **19**, 103–6.

Craig, J. (1985) Tectonic Evolution of the Area Between Borth and Cardigan, Dyfed, West Wales. Unpublished PhD thesis, University of Wales, Aberystwyth.

Craig, J. (1987) The structure of the Llangranog Lineament, West Wales: a Caledonian transpression zone. In *Sedimentation and Tectonics of the Welsh Basin.* (eds W.R. Fitches and N.H. Woodcock), Geological Journal, 22, 167–81.

Craig, J., Fitches, W.R., and Maltman, A.J. (1982) Chlorite-mica stacks in low-strain rocks from central Wales. *Geological Magazine*, **119**, 243–56.

Davies, W. and Cave, R. (1976) Folding and cleavage determined during sedimentation. *Sedimentary Geology*, **15**, 89–134.

Dearman, W.R., Shiels, K.A.G., and Larwood, G.P. (1962). Refolded folds in the Silurian rocks of Eyemouth, Berwickshire. *Proceedings of the Yorkshire Geological Society*, **33**, 273–86.

Dewey, J.F. (1969) Structure and sequence in paratectonic British Caledonides. In *North Atlantic – Geology and Continental Drift.* (ed. M. Kay), Memoir of the American Association of Petroleum Geologists, 12, 309–35.

Dickson, J.A.D. (1983) Graphic modelling of crystal aggregates and its relevance to cement diagenesis. *Philosophical Transactions of the Royal Society of London*, **A309**, 465–502.

Downie, C. and Soper, N.J. (1972) Age of the Eycott Volcanic Group and its conformable relationship to the Skiddaw Slates in the English Lake District. *Geological Magazine*, **109**, 259–68.

Eastwood, T., Hollingworth, S.E., Rose, W.C.C *et al.* (1968) *The Geology of the Country around Cockermouth*. Memoir of the Geological Survey of the UK, HMSO, London.

Elles, G.L. (1940) The stratigraphy and faunal succession in the Ordovician rocks of the Builth–Llandrindod inlier, Radnorshire. *Quarterly Journal of the Geological Society of London*, **95**, 385–445.

Fearnsides, W.G. (1910) The Tremadoc Slates and associated rocks of south-east Caernarvonshire. *Quarterly Journal of the Geological Society of London*, **66**, 142–88.

Fearnsides, W.G. and Davies, W. (1944) The geology of Deudraeth – the country between Traeth Mawr and Traeth Bach, Merioneth. *Quarterly Journal of the Geological Society of London*, **99**, 247–76.

Firman, R.J. (1957) The Borrowdale Volcanic Series between Wastwater and the Duddon Valley, Cumberland. *Proceedings of the Yorkshire Geological Society*, **31**, 39–64.

Firman, R.J. and Lee, M.K. (1986) Age and structure of the concealed English Lake District batholith and its probable influence on subsequent sedimentation, tectonics and mineralization. In *Geology in the Real World – the Kingsley Dunham volume.* (eds R.W. Nesbitt and I. Nichol), Institution of Mining and Metallurgy, London, pp. 117–27.

Fitch, F.J., Miller, J.A., Evans, A.L., *et al.* (1969) Isotopic age determinations on rocks from Wales and the Welsh Borders. In *The Precambrian and Lower Palaeozoic Rocks of Wales.* (ed. A. Wood), University of Wales Press, Cardiff, pp. 23–46.

Fitches, W.R. (1972). Polyphase deformation structures in the Welsh Caledonides near Aberystwyth. *Geological Magazine*, **109**, 149–55.

Fitches, W.R. (1987) Aspects of veining in the Lower Palaeozoic rocks of the Welsh Basin. In *Deformation of Sediments and Sedimentary Rocks.* (eds M.R. Jones and R.M.F. Preston), Geological Society of London Special Publication, No 29, pp. 297–317.

References

Fitches, W.R. and Campbell, S.D.G. (1987) Tectonic evolution of the Bala Lineament in the Welsh Basin. In *Sedimentation and Tectonics of the Welsh Basin* (eds W.R. Fitches and N.H. Woodcock), Geological Journal, **22**, 131–53.

Fitches, W.R. and Johnson, R. (1978) Cleavage–fold relationships in the Aberystwyth Grits – a preliminary report (Abstract of Tectonic Studies Group conference report: Deformation of soft sediments, Aberystwyth, March 1977), *Journal of the Geological Society of London*, **135**, 250.

Fitches, W.R., Cave, R., Craig, J. *et al.* (1986) Early veins as evidence of detachment structures in the Lower Palaeozoic rocks of the Welsh Lower Palaeozoic Basin. *Journal of Structural Geology*, **8**, 607–20.

Fitton, J.G. and Hughes, D.J. (1970) Volcanism and plate tectonics in the British Ordovician. *Earth and Planetary Science Letters*, **8**, 223–8.

Fortey, N.J. (1989) Low grade metamorphism in the Lower Ordovician Skiddaw Group of the Lake District, England. *Proceedings of the Yorkshire Geological Society*, **47**, 325–37.

Francis, E.H. and Howells, M.F. (1973) Transgressive welded ash-flow tuffs among the Ordovician sediments of NE Snowdonia, N. Wales. *Journal of the Geological Society of London*, **129**, 621–41.

Furnes, H. (1978) A Comparative Study of Caledonian Volcanics in Wales and West Norway. Unpublished PhD thesis, University of Oxford.

Fyfe, T.B. and Weir, J.A. (1976) The Ettrick Valley Thrust and the upper limit of the Moffat Shales in Craigmichan Scaurs. *Scottish Journal of Geology*, **12**, 93–102.

Garwood, E.J. and Goodyear, E. (1918) On the geology of the Old Radnor district, with special reference to an algal development in the Woolhope Limestone. *Quarterly Journal of the Geological Society of London*, **74**, 1–30.

Geikie, A. (1863) *The geology of eastern Berwickshire*. Memoir of the Geological Survey of Scotland, 58 pp.

George, T.N. (1961) *British Regional Geology, North Wales*. (3rd edn), HMSO 97 pp.

George, T.N. (1963) Palaeozoic growth of the British Caledonides. In *The British Caledonides*. (eds M.R.J. Johnson, and F.H. Stewart), Oliver and Boyd, Edinburgh, pp. 1–33.

Gibbons, W. (1983) The Monian 'Penmynydd Zone of metamorphism' in Llyn, North Wales. *Geological Journal*, **18**, 21–41.

Gibbons, W. (1987) Menai Straits fault system: an early Caledonian terrane boundary in north Wales. *Geology*, **15**, 744–7.

Green, J.F.N. (1915) The structure of the eastern part of the Lake District. *Proceedings of the Geologists' Association*, **26**, 195–223.

Green, J.F.N. (1917) The age of the chief intrusions of the Lake District. *Proceedings of the Geologists' Association*, **28**, 1–30.

Green, J.F.N. (1920) The geological structure of the Lake District. *Proceedings of the Geologists' Association*, **31**, 109–26.

Greenly, E. (1919) *The Geology of Anglesey*. Memoir of the Geological Survey of the UK (2 volumes), HMSO, London.

Greenly, E. (1920) One inch Geological Map of Anglesey (Sheet 92). Geological Survey of the UK, HMSO, London.

Greig, D.E., Wright, J.E., Hains, B.A., and Mitchell, G.H. (1968) *Geology of the country around Church Stretton, Craven Arms, Wenlock Edge and Brown Clee*. (Sheet 166), Memoir of the Geological Survey of Great Britain, HMSO, London.

Grigor'ev, D.P. (1965) The ontogeny of minerals. *Israeli Program for Scientific Translation*.

Hall, J. (1815) On the vertical position and convolutions of certain strata and their relation with granite. *Transactions of the Royal Society of Edinburgh: Earth Sciences*, 7, 79–108.

Hamdi Lemnouar, D.E.S. (1988) Relationship between Mineral Deposition and Movement in Fault Zones from Llangollen, North Wales. Unpublished MSc thesis, University of Manchester.

Harland, W.B. (1971) Tectonic transpression in Caledonian Spitsbergen. *Geological Magazine*, **108**, 27–42.

Helm, D.G. (1970) Stratigraphy and structure in the Black Combe Inlier, English Lake District. *Proceedings of the Yorkshire Geological Society*, **38**, 105–48.

Helm, D.G. and Roberts, B. (1971) The relationship between the Skiddaw and Borrowdale Volcanic Groups in the English Lake District. *Nature*, **232**, 181–3.

Helm, D.G. and Siddans, A.W.B. (1971) Deformation of a slaty, lapillar tuff in the English Lake District, discussion. *Bulletin of the Geological Society of America*, **82**, 523–31.

Helm, D.G., Roberts, B., and Simpson, A. (1963) Polyphase folding in the Caledonides south of the Scottish Highlands. *Nature*, **200**, 1060–2.

Hollingworth, S.E. (1955) The geology of the Lake District – a review. *Proceedings of the Geologists'*

References

Association, **65**, 385–402.

Howells, M.F., Francis, E.H., Leveridge, B.E., and Evans C.D.R. (1978) *Capel Curig and Betws-y-Coed. (Description of 1:25 000 sheet SH 75) Classical areas of British Geology*, Institute of Geological Sciences, HMSO, London, 73 pp.

Howells, M.F., Leveridge, B.E., Addison, R., *et al.* (1979) The Capel Curig Volcanic Formation, Snowdonia, North Wales; variations in ash-flow tuffs related to emplacement environment. In *The Caledonides of the British Isles – Reviewed.* (eds A.L. Harris, C.H. Holland, and B.E. Leake) Special Publication of the Geological Society of London, No. 8, pp. 611–18.

Howells, M.F., Leveridge, B.E., and Reedman, A.J. (1981). *Snowdonia. (Rocks and fossils Series, No. 1*, Unwin paperbacks), London, 119 pp.

Howells, M.F., Reedman, A.J., and Campbell, S.D.G. (1986) The submarine eruption and emplacement of the .Lower Rhyolitic Tuff Formation (Ordovician), N. Wales. *Journal of the Geological Society of London*, **143**, 411–23.

Hutton, D.H.W. (1982) A tectonic model for the emplacement of the Main Donegal Granite, NW Ireland. *Journal of the Geological Society of London*, **139**, 615–31.

Hutton, D.H.W. (1987) Strike-slip terranes and a model for the evolution of the British and Irish Caledonides. *Geological Magazine*, **124**, 405–25.

Hutton, D.H.W. and Murphy, F.C. (1987) The Silurian of the Southern Uplands and Ireland as a successor basin to the end-Ordovician closure of Iapetus. *Journal of the Geological Society of London*, **144**, 765–72.

Hutton, J. (1795) *Theory of the Earth 1.* Creech, Edinburgh.

Institute of Geological Sciences (1977) *Llandrindod Wells Ordovician inlier* (1:25,000 geological map). HMSO, London.

Jackson, D.E. (1961) Stratigraphy of the Skiddaw Group between Buttermere and Mungrisdale, Cumberland. *Geological Magazine*, **98**, 515–28.

Jackson, D.E. (1962) Graptolite zones in the Skiddaw Group in Cumberland, England. *Journal of Palaeontology*, **36**, 300–13.

Jackson, D.E. (1978) The Skiddaw Group. In *The Geology of the Lake District.* (ed. F. Moseley), Yorkshire Geological Society Occasional Publication, No. 3, pp. 79–98.

Jeans, P.J.F. (1971) The relationship between the Skiddaw Slates and the Borrowdale Volcanics. *Nature, Physical Science*, **234**, 59.

Jeans, P.J.F. (1972) The junction between the Skiddaw Slates and Borrowdale Volcanics in Newlands Beck, Cumberland. *Geological Magazine*, **109**, 25–8.

Jeans, P.J.F. (1974) The Structure, Metamorphism and Stratigraphy of the Skiddaw Slates East of Crummock Water, Cumberland. Unpublished PhD thesis, University of Birmingham.

Jennings, A.V. and Williams, G.J. (1891) Manod and the Moelwyns. *Quarterly Journal of the Geological Society of London*, **47**, 368–83.

Johnson, M.R.W., Sanderson, D.J., and Soper, N.J. (1979) Deformation in the Caledonides of England, Ireland and Scotland. In *The Caledonides of the British Isles – Reviewed.* (eds A.L. Harris, C.W. Holland, and B.E. Leake), Geological Society of London Special Publication, No. 8, pp. 165–86.

Jones, O.T. (1912) The geological structure of Central Wales and the adjoining regions. *Quarterly Journal of the Geological Society of London*, **68**, 328–73.

Jones, O.T. (1954) The trends of geological structures in relation to directions of maximum compression. *British Association for the Advancement of Science, London*, **11**, 102–6.

Jones, O.T. (1955) The geological evolution of Wales and the adjacent regions. *Quarterly Journal of the Geological Society of London*, **111**, 323–51.

Jones, O.T. and Pugh, W.J. (1915) The geology of the district around Machynlleth and the Llyfnant Valley. *Quarterly Journal of the Geological Society of London*, **71**, 343–85.

Jones, O.T. and Pugh, W.J. (1941) The Ordovician rocks of the Builth District; a preliminary account. *Geological Magazine*, **78**, 185–91.

Jones, O.T. and Pugh, W.J. (1946) The complex intrusion of Welfield, near Builth Wells, Radnorshire. *Quarterly Journal of the Geological Society of London*, **102**, 157–88.

Jones, O.T. and Pugh, W.J. (1949) An early Ordovician shoreline in Radnorshire, near Builth Wells. *Quarterly Journal of the Geological Society of London*, **105**, 65–99.

Kelling, G. (1961) The stratigraphy and structure of the Ordovician rocks of the Rhinns of Galloway. *Quarterly Journal of the Geological Society of London*, **117**, 37–75.

Kemp, A.E.S. (1986) Tectonostratigraphy of the Southern Belt of the Southern Uplands. *Scottish Journal of Geology*, **22**, 241–56.

Kemp, A.E.S. (1987) Tectonic development of the Southern Belt of the Southern Uplands

References

accretionary complex. *Journal of the Geological Society of London*, **144**, 827–38.

Kemp, A.E.S. and White, D.E. (1985) Silurian trench sedimentation in the Southern Uplands, Scotland: implications of new age data. *Geological Magazine*, **122**, 275-7.

King, W.B.R. and Wilcockson, W.H. (1934) The Lower Palaeozoic rocks of Austwick and Horton-in-Ribblesdale, Yorkshire. *Quarterly Journal of the Geological Society of London*, **90**, 7–31.

Kirk, N.H. (1951) The upper Llandovery and lower Wenlock rocks of the area between Dolyhir and Presteigne, Radnorshire. *Proceedings of the Geological Society of London*, **1471**, 56–8.

Kirk, N.H. (1952) The tectonic structure of the anticlinal disturbance of Breconshire and Radnorshire: Port Faen to Presteigne. *Proceedings of the Geological Society of London*, **1485**, 87–91.

Knipe, R.J. and Needham, D.T. (1986) Deformation processes in accretionary wedges – examples from the SW margin of the Southern Uplands, Scotland. In *Collision Tectonics*. (eds M.P. Coward and A.I. Ries), Geological Society of London Special Publication, No. 19, pp. 53–67.

Knipe, R.J. and White, S.H. (1977) Microstructural variation of an axial plane cleavage around a fold – a H.V.E.M. study. *Tectonophysics*, **39**, 355–80.

Kokelaar, B.P. (1977) The Igneous History of the Rhobell Fawr Area, Merionethshire, North Wales. Unpublished PhD thesis, University of Wales, Aberystwyth.

Kokelaar, B.P. (1979) Tremadoc to Llanvirn volcanism on the south-east side of the Harlech Dome (Rhobell Fawr), North Wales. In *The Caledonides of the British Isles – reviewed*. (eds A.L. Harris, C.H. Holland, and B.E. Leake). Geological Society of London Special Publication, No. 8, pp. 591–6.

Kokelaar, B.P. (1988) Tectonic controls of Ordovician arc and marginal basin volcanism in Wales. *Journal of the Geological Society of London*, **145**, 759–75.

Lapworth, C. (1874) On the Silurian rocks of the south of Scotland. *Transactions of the Geological Society of Glasgow*, **4**, 164–74.

Lapworth, C. (1889) On the Ballantrae rocks of South Scotland. *Geological Magazine*, **26**, 20–4, 59–69.

Lawrence, D.J.D., Webb, B.C., Young, B. (1986) *The Geology of the Late Ordovician and Silurian Rocks (Windermere Group) in the Area Around Kentmere and Crook*. Report of the British Geological Survey, Volume 18, No. 5.

Leake, B.E., Tanner, P.W.C., Singh, D.,and Halliday, A.N. (1983) Major southward thrusting of the Dalradian rocks of Connemara, W. Ireland. *Nature*, **305**, 210–13.

Leggett, J.K. (1980) The sedimentological evolution of a Lower Palaeozoic accretionary fore-arc in the Southern Uplands of Scotland. *Sedimentology*, **27**, 401–17.

Leggett, J.K., McKerrow, W.S., and Eales, M.H. (1979) The Southern Uplands of Scotland: a Lower Palaeozoic accretionary prism. *Journal of the Geological Society of London*, **136**, 755–70.

Leggett, J.K., McKerrow, W.S., and Soper, N.J. (1983) A model for the crustal evolution of southern Scotland. *Tectonics*, **2**, 187–210.

Lisle, R.J. (1977) Clastic grain shape and orientation in relation to cleavage from the Aberystwyth Grits, Wales. *Tectonophysics*, **39**, 381–95.

Lynas, B.D.T. (1970) Clarification of the polyphase deformations of North Wales Palaeozoic rocks. *Geological Magazine*, **107**, 505–10.

Lynas, B.D.T. (1973) The Cambrian and Ordovician rocks of the Migneint area, North Wales. *Journal of the Geological Society of London*, **129**, 481–503.

Lynas, B.D.T. (1988) Evidence for dextral oblique-slip faulting in the Shelve Ordovician Inlier, Welsh Borderland: implications for the south British Caledonides. *Geological Journal*, **23**, 39–57.

Marr, J.E. (1916) *The Geology of the Lake District*, Cambridge University Press, Cambridge, 220 pp.

Martin, B.A., Howells, M.F., and Reedman, A.J. (1981) A geological reconnaissance study of the Dyfi Valley region, Gwynedd/Powys, Wales. *Institute of Geological Sciences Environmental Protection Unit*, **81/1**, 57 pp.

Matley, C.A. (1899) On the geology of northern Anglesey (with appendix by W.W. Watts). *Quarterly Journal of the Geological Society of London*, **55**, 635–80.

Matley, C.A. (1901) The geology of the Mynydd-y-Garn (Anglesey). *Quarterly Journal of the Geological Society of London*, **57**, 20–30.

McCabe, P.J. (1972) The Wenlock and Lower Ludlow strata of the Austwick and Horton-in-Ribblesdale Inlier of north-west Yorkshire. *Proceedings of the Yorkshire Geological Society*, **39**, 167–74.

McCabe, P.J. and Waugh, B. (1983) Wenlock and

Ludlow sedimentation in the Austwick and Horton-in-Ribblesdale Inlier, north-west Yorkshire, *Proceedings of the Yorkshire Geological Society*, **39**, 445–70.

McKerrow, W.S. (1988) Wenlock to Givetian deformation in the British Isles and the Canadian Appalachians. In *The Caledonian–Appalachian Orogen*. (eds A.L. Harris and D.J. Fettes), Geological Society Special Publication, No. 38, 405–12.

McKerrow, W.S. and Cocks, L.R.M. (1976) Progressive faunal migration across the Iapetus Ocean. *Nature*, **263**, 304-6.

McKerrow, W.S. and Soper, N.J. (1989) The Iapetus Suture in the British Isles. *Geological Magazine*, **126**, 1-8.

McKerrow, W.S., Leggett, J.K., and Eales, M.H. (1977) Imbricate thrust model of the Southern Uplands of Scotland. *Nature*, **267**, 237–9.

Mitchell, A.H.G. and McKerrow, W.S. (1975) Analogous evolution of the Burma Orogen and the Scottish Caledonides. *Bulletin of the Geological Society of America*, **86**, 305–15.

Mitchell, G.H. (1929) The succession and structure of the Borrowdale Volcanic Series in Troutbeck, Kentmere and the western part of Long Sleddale (Westmorland). *Quarterly Journal of the Geological Society of London*, **85**, 9–43.

Mitchell, G.H. (1940) The Borrowdale Volcanic Series of Coniston, Lancashire. *Quarterly Journal of the Geological Society of London*, **96**, 301–19.

Mitchell, G.H. (1956a) The geological history of the Lake District. *Proceedings of the Yorkshire Geological Society*, **30**, 407–63.

Mitchell, G.H. (1956b) The Borrowdale Volcanic Series of the Dunnerdale Fells, Lancashire. *Liverpool and Manchester Geological Journal*, **1**, 428–49.

Mitchell, G.H., Moseley, F., Firman, R.J., *et al.* (1972) Excursion to the northern Lake District. *Proceedings of the Geologists' Association*, **83**, 443–70.

Moore, J.G. and Peck, D.L. (1962) Accretionary lapilli in volcanic rocks of the Western Continental United States. *Journal of Geology, Chicago*, **70**, 182–93.

Morris, J.H. (1987) The Northern Belt of the Longford-Down Inlier, Ireland and Southern Uplands, Scotland: an Ordovician back-arc basin. *Journal of the Geological Society of London*, **144**, 773–86.

Morris, T.O. and Fearnsides, W.G. (1926). The stratigraphy and structure of the Cambrian slate belt of Nantlle (Caernarvonshire). *Quarterly Journal of the Geological Society of London*, **82**, 250–303.

Moseley, F. (1964) The succession and structure of the Borrowdale Volcanic rocks north-west of Ullswater. *Geological Journal*, **4**, 127–42.

Moseley, F. (1968) Joints and other structures in the Silurian rocks of the southern Shap Fells, Westmorland. *Geological Journal*, **6**, 79–96.

Moseley, F. (1972) A tectonic history of north-west England. *Journal of the Geological Society of London*, **128**, 561–98.

Moseley, F. (1975) Structural relations between the Skiddaw Slates and the Borrowdale Volcanics. *Proceedings of the Cumberland Geological Society*, **3**, 127–45.

Moseley, F. (1977) Caledonian plate tectonics and the place of the English Lake District. *Bulletin of the Geological Society of America*, **88**, 764–8.

Moseley, F. (1981) *Methods in Field Geology*. Freeman, 211 pp.

Moseley, F. (1983) *The Volcanic Rocks of the Lake District*. Macmillan, London. 111 pp.

Moseley, F. (1984) Lower Palaeozoic lithostratigraphical classification in the English Lake District. *Geological Journal*, **19**, 239–47.

Moseley, F. (1986) *Geology and Scenery in the Lake District*. Macmillan, London, 88 pp.

Moseley, F. (1990) Geology of the Lake District. *Geologists' Association Guide*, Geologists' Association, London, 1–213 pp.

Muir, M.D., Bliss, G.M., Grant, P.R., and Fisher, M. (1979) Palaeontological evidence for the age of some supposedly Precambrian rocks in Anglesey, North Wales. *Journal of the Geological Society of London*, **136**, 61–4.

Mukhopadhyay, D. (1972) Deformation of a slaty, lapillar tuff in the Lake District, England: discussion. *Bulletin of the Geological Society of America*, **83**, 547–8.

Murphy, F.C. (1985) Non-axial planar cleavage and Caledonian sinistral transpression in eastern Ireland. *Geological Journal*, **20**, 257–79.

Murphy, F.C. and Hutton, D.H.W. (1986) Is the Southern Uplands of Scotland really an accretionary prism? *Geology*, **14**, 354–7.

Nettle, J.T. (1964) Fabric analysis of a deformed vein. *Geological Magazine*, **101**, 220–7.

Nicholson, R. (1966) The problem of origin, deformation and recrystallization of calcite-quartz bodies. *Geological Journal*, **5**, 117–26.

Nicholson, R. (1970) Tectonic ripples and associated minor structures in the Silurian rocks of Denbighshire. *Geological Magazine*, **107**, 447–9.

Nicholson, R. (1978) Folding and pressure solution in a laminated calcite–quartz vein from the Silurian slates of the Llangollen region of North Wales. *Geological Magazine*, **115**, 47–54.

Norman, T.N. (1961) The Geology of the Silurian Strata in the Blawith Area, Furness. Unpublished PhD thesis, University of Birmingham.

Numan, N.M.S. (1974) Structure and Stratigraphy of the Southern Part of the Borrowdale Volcanic Group, English Lake District. Unpublished PhD thesis, University of Sheffield.

Nutt, M.J.C. and Smith, E.G. (1981) Transcurrent faulting and the anomalous position of pre-Carboniferous Anglesey. *Nature*, **290**, 492–5.

Oertel, G. (1970) Deformation of a slaty, lapillar tuff in the Lake District, England. *Bulletin of the Geological Society of America*, **81**, 1173–87.

Oertel, G. (1971) Deformation of a slaty, lapillar tuff in the English Lake District: reply. *Bulletin of the Geological Society of America*, **82**, 533–6.

Oertel, G. (1972) Deformation of a slaty, lapillar tuff in the Lake District, England: another reply. *Bulletin of the Geological Society of America*, **83**, 549–50.

Oertel, G. and Wood, D.S. (1974) Finite strain measurement: a comparison of methods. *Transactions of the American Geophysical Union*, **55**, 695.

Orton, G. (1988) Volcanoclastic Sedimentation in a Caradocian Marginal Basin, North Wales. Unpublished PhD thesis, University of Oxford.

Owen, T.R. (1974) The Variscan Orogeny in Wales. In *The Precambrian and Lower Palaeozoic rocks of Wales*. (ed. A. Wood), University of Wales Press, Cardiff, pp. 285–94.

Owen, T.R. and Weaver, J.D. (1983) The structure of the main South Wales coalfield and its margins. In *The Variscan Fold Belt in the British Isles*. (ed. P.L. Hancock), Adam Hilger Ltd, Bristol, pp. 74–87.

Peach, B.N. and Horne, J. (1899) *The Silurian rocks of Britain I: Scotland*. Memoir of the Geological Survey of the UK, HMSO, London.

Phillips, W.J. (1972) Hydraulic fracturing and mineralization. *Journal of the Geological Society of London*, **128**, 337–59.

Phillips, W.E.A., Stillman, C.J., and Murphy, T. (1976) A Caledonian plate tectonic model. *Journal of the Geological Society of London*, **132**, 579-609.

Playfair, J. (1805) Biographical account of the late Dr James Hutton. *Transactions of the Royal Society of Edinburgh*, **5**, 39–99.

Powell, C. McA. (1979) A morphological classification of rock cleavage. *Tectonophysics*, **58**, 21–34.

Powell, D. and Phillips, W.E.A. (1985) Time of deformation in the Caledonide Orogen of Britain and Ireland. In (ed. A.L. Harris), *Geological Society of London, Memoir*, **9**, pp. 17–39.

Price, N.J. (1958) Tectonics of the Aberystwyth Grits. Unpublished PhD thesis, University of Wales, Aberystwyth.

Price, N.J. (1962) The tectonics of the Aberystwyth grits. *Geological Magazine*, **99**, 542–57.

Pringle, J. and George, T.N. (1948) *British Regional Geology South Wales*, 2nd edn, HMSO, 152 pp.

Ramsay, A.C. (1866) *The Geology of North Wales*, 1st edn, Memoir of the Geological Survey of the UK, HMSO, London.

Ramsay, A.C. (1881) *The Geology of North Wales*, 2nd edn, Memoir of the Geological Survey of the UK, HMSO, London.

Ramsay, J.G. (1967) *Folding and Fracturing of rocks*. McGraw-Hill, New York, 568 pp.

Ramsay, J.G. (1980) The crack–seal mechanism of rock deformation. *Nature*, **284**, 135–9.

Ramsay, J.G. (1983) Rock ductility and its influence on the development of tectonic structures of mountain belts. In *Mountain Building Processes*. (ed. K.J. Hsü), Academic Press, London, pp. 111–27.

Ramsay, J.G. and Huber, M.I. (1983) *The Techniques of Modern Structural Geology (Volume 1: Strain Analysis)*. Academic Press, London.

Ramsay, J.G. and Huber, M.I. (1987) *The Techniques of Modern Structural Geology (Volume 2: Folds and Fractures)*. Academic Press, London.

Ramsay, J.G. and Wood, D.S. (1973) The geometric effects of volume change during deformational processes. *Tectonophysics*, **16**, 263–77.

Rast, N. (1969) The relationship between Ordovician structure and volcanicity in Wales. In *The Precambrian and Lower Palaeozoic rocks of Wales*. (ed. A. Wood), University of Wales Press, Cardiff, pp. 305–36.

Raybould, J.G. (1976) The influence of pre-existing planes of weakness in rocks on the localization of vein-type ore deposits. *Economic Geologist*, **71**, 636–41.

Reedman, A.J., Colman, T.B., Campbell, S.D.G., and Howells, M.F. (1985) Volcanogenic mineralization related to the Snowdon Volcanic Group (Ordovician), North Wales. *Journal of the Geological Society of London*, **142**, 875–88.

Roberts, B. (1967) Succession and structure in

Llwyd Mawr Syncline, Caernarvonshire North Wales. *Geological Journal*, **5**, 369–90.

Roberts, B. (1969) The Llwyd Mawr ignimbrite and its associated volcanic rocks. In *The Precambrian and Lower Palaeozoic rocks of Wales*. (ed. A. Wood), University of Wales Press, Cardiff, pp. 337–56.

Roberts, B. (1979) *The Geology of Snowdonia and Llyn: An outline and field guide*. Adam Hilger Ltd, Bristol, 183 pp.

Roberts, B. and Siddans, A.W.B. (1971) Fabric studies in the Llwyd Mawr ignimbrite, Caernarfonshire, North Wales. *Tectonophysics*, **12**, 283–306.

Roberts, D.E. (1971). Structures of the Skiddaw Slates in the Caldew Valley, Cumberland. *Geological Journal*, **7**, 225–38.

Roberts, D.E. (1973) The structure of the Skiddaw Slates in the Blencartha–Mungrisdale Area, Cumberland. Unpublished PhD thesis, University of Birmingham.

Roberts, D.E. (1977a) The structure of the Skiddaw Slates in the Blencathra–Mungrisdale area, Cumbria. *Geological Journal*, **12**, 33–58.

Roberts, D.E. (1977b) Minor tectonic structures in the Skiddaw Slates of Raven Crags, Mungrisdale, northern Lake District. *Proceedings of the Geologists' Association*, **88**, 117–24.

Roberts, J.L. and Sanderson, D.J. (1974) Oblique fold axes in the Dalradian rocks of the southwest Highlands. *Scottish Journal of Geology*, **9**, 281–96.

Rock, N.M.S., Gaskarth, J.W., and Rundle, C.C. (1986) Late Caledonian dyke-swarms in Southern Scotland: a regional zone of K-rich lamprophyres and associated vents. *Journal of Geology*, **94**, 505-22.

Rose, W.C.C. (1954) The sequence and structure of the Skiddaw Slates in the Keswick–Buttermere area. *Proceedings of the Geologists' Association*, **65**, 403–6.

Rundle, C.C. (1981) The significance of isotopic dates from the English Lake District for the Ordovician–Silurian time scale. *Journal of the Geological Society of London*, **138**, 569–72.

Rust, B.R. (1965) The stratigraphy and structure of the Whithorn area of Wigtownshire, Scotland. *Scottish Journal of Geology*, **1**, 101–33.

Sanderson, D.J. and Marchini, W.R.D. (1984) Transpression. *Journal of Structural Geology*, **6**, 449–58.

Sanderson, D.J., Andrews, J.R., Phillips, W.E.A., and Hutton, D.H.W. (1980) Deformation studies in the Irish Caledonides. *Journal of the Geological Society of London*, **137**, 289–302.

Schaller, W.T. (1932) The crystal cavities of the New Jersey zeolite region, *United States Geological Survey Bulletin*, **32**, 47–58.

Shackleton, R.M. (1953) The structural evolution of North Wales. *Liverpool and Manchester Geological Journal*, **1**, 261–97.

Shackleton, R.M. (1954) The structure and succession of Anglesey and the Lleyn Peninsula. *British Association for the Advancement of Science*, **11**, 106–8.

Shackleton, R.M. (1959) The stratigraphy of the Moel Hebog district between Snowdon and Tremadoc. *Liverpool and Manchester Geological Journal*, **2**, 216–52.

Shackleton, R.M. (1969) The Pre-Cambrian of North Wales. In *The Precambrian and Lower Palaeozoic rocks of Wales*. (ed. A. Wood), University of Wales Press, Cardiff, pp. 1–18.

Sharp, D. (1849) On slaty cleavage. *Quarterly Journal of the Geological Society of London*, **5**, 111–15.

Shepherd, T.J., Beckinsale, R.D., Rundle, C.C. *et al.* (1976) Genesis of the Carrock Fell tungsten deposits, Cumbria: fluid inclusion and isotopic study. *Transactions of the Institution of Mining and Metallurgy*, **B85**, 63–73.

Siddans, A.W.B. (1971) The origin of slaty cleavage. Unpublished PhD thesis, University of London.

Siddans, A.W.B. (1972) Slaty cleavage – a review of research since 1815. *Earth Science Reviews*, **8**, 205–32.

Simpson, A. (1967) The stratigraphy and tectonics of the Skiddaw Slates and the relationship of the overlying Borrowdale Volcanic Series in part of the Lake District. *Geological Journal*, **5**, 391–418.

Smith, M. (1987) The Tremadoc 'Thrust' Zone in southern central Snowdonia. In *Sedimentation and Tectonics of the Welsh Basin*. (eds W.R. Fitches and N.H. Woodcock), Geological Journal, 22, 119–29.

Smith, M. (1988) The Tectonic Evolution of the Cambro-Ordovician Rocks of Southern Central Snowdonia. Unpublished PhD thesis, University of Wales, Aberystwyth.

Sokoutis, D. (1987) Finite strain effects in experimental mullions. *Journal of Structural Geology*, **9**, 233–42.

Soper, N.J. (1970) Three critical localities on the junction of the Borrowdale Volcanic Rocks with the Skiddaw Slates in the Lake District. *Proceedings of the Yorkshire Geological Society*, **37**, 461–93.

Soper, N.J. (1986) The Newer Granite problem: a geotectonic view. *Geological Magazine*, **123**, 227–36.

Soper, N.J. (1987) The Ordovician batholith of the English Lake District. *Geological Magazine*, **124**, 481–2.

Soper, N.J. (1988) Timing and geometry of collision, terrane accretion and sinistral strike-slip events in the British Caledonides. In *The Caledonian–Appalachian Orogen*. (eds A.L. Harris and D.J. Fettes), Geological Society Special Publication, No. 38, pp. 482–92.

Soper, N.J. and Hutton, D.H.W. (1984) Late Caledonian sinistral displacements in Britain: implications for a three-plate collision model. *Tectonics*, **3**, 781–94.

Soper, N.J. and Moseley, F. (1978) Structure. In *The Geology of the Lake District*. (ed. F. Moseley), Yorkshire Geological Society Occasional Publication, No. 3, pp. 45–67.

Soper, N.J. and Numan, N.M.S. (1974) Structure and stratigraphy of the Borrowdale Volcanic rocks of the Kentmere area, English Lake District. *Geological Journal*, **9**, 147–66.

Soper, N.J. and Roberts, D.E. (1971) Age of cleavage in the Skiddaw Slates in relation to the Skiddaw aureole. *Geological Magazine*, **108**, 293–302.

Soper, N.J., Webb, B.C., and Woodcock, N.H. (1987). Late Caledonian (Acadian) transpression in north-west England: timing, geometry and geotectonic significance. *Proceedings of the Yorkshire Geological Society*, **46**, 175–92.

Sorby, H.C. (1853) On the origin of slaty cleavage. *New Philosophical Journal of Edinburgh*, **55**, 137–48.

Sorby, H.C. (1856) On slaty cleavage as exhibited in the Devonian limestones of Devonshire. *Philosophical Magazine*, **11**, 20–37.

Sorby, H.C. (1908) On the application of quantitative methods to the study of the structure and history of rocks. *Quarterly Journal of the Geological Society of London*, **64**, 171–232.

Stone, P., Floyd, J.D., Barnes, R.P., and Lintern, B.C. (1987). A sequential back-arc and foreland basin thrust duplex model for the Southern Uplands of Scotland. *Journal of the Geological Society of London*, **144**, 753–64.

Stringer, P. and Treagus, J.E. (1980) Non-axial planar S_1 cleavage in the Hawick Rocks of the Galloway area, Southern Uplands, Scotland. *Journal of Structural Geology*, **2**, 317–31.

Stringer, P. and Treagus, J.E. (1981). Asymmetrical folding in the Hawick Rocks of the Galloway area, Southern Uplands. *Scottish Journal of Geology*, **17**, 129–48

Stringer, P. and Treagus, J.E. (1983) Asymmetrical folding in the Hawick Rocks of the Galloway area, Southern Uplands: Reply. *Scottish Journal of Geology*, **19**, 107–12.

Thirlwall, M.F. (1988) Geochronology of Late Caledonian magmatism in northern Britain. *Journal of the Geological Society of London*, **145**, 951–67.

Toghill, P. (1970) The south-east limit of the Moffat Shales in the upper Ettrick Valley region, Selkirkshire. *Scottish Journal of Geology*, **6**, 233–42.

Treagus, J.E. and Treagus, S.H. (1981) Folds and the strain ellipsoid:- a general model. *Journal of Structural Geology*, **3**, 1–18.

Tremlett, C.R. (1982) The structure and structural history of the Lower Palaeozoic Rocks in parts of North Dyfed, Wales. Unpublished PhD thesis, University of Wales, Aberystwyth.

Tullis, T.E. and Wood, D.S. (1972) The relationship between preferred orientation and finite strain for three slates. *Geological Society of America (Abstract with program)*, **4**, (**7**), p. 694.

Tullis, T.E. and Wood, D.S. (1975) Correlation of finite strain from both reduction bodies and preferred orientation of mica in slate from Wales. *Bulletin of the Geological Society of America*, **86**, 632–8.

Tyler, J.E. and Woodcock, N.H. (1987) The Bailey Hill Formation: Ludlow Series turbidites in Welsh Borderland reinterpreted as distal storm deposits. *Geological Journal*, **22**, 73–86.

Wadge, A.J. (1972) Sections through the Skiddaw–Borrowdale unconformity in eastern Lakeland. *Proceedings of the Yorkshire Geological Society*, **39**, 179–98.

Wadge, A.J. (1978a) Classification and stratigraphical relationships of the Lower Ordovician rocks. In *The Geology of the Lake District*. (ed. F. Moseley), Yorkshire Geological Society Occasional Publication No. 3, pp. 68–78.

Wadge, A.J. (1978b). Devonian. In *The Geology of the Lake District*. (ed. F. Moseley), Yorkshire Geological Society Occasional Publication No. 3, pp. 164–7.

Wadge, A.J., Gale, N.H., Beckinsale, R.D., and Rundle, C.C. (1978) A Rb–Sr isochron for the Shap granite. *Proceedings of the Yorkshire Geological Society*, **42**, 297–305.

Walton, E.K. (1961) Some aspects of the succession and structure in the Lower Palaeozoic rocks of the Southern Uplands of Scotland. *Geologische Rundschaft*, **50**, 63–77.

Walton, E.K. (1965) Lower Palaeozoic rocks: stratigraphy, palaeogeography and structure. In *Geology of Scotland*. 1st edn, (ed. G.Y. Craig), Oliver and Boyd, Edinburgh, pp. 167–227.

Walton, E.K. (1983) Lower Palaeozoic rocks, structure and palaeogeography. In *Geology of Scotland*. 2nd edn, (ed. G.Y. Craig), Edinburgh, pp. 139–66.

Ward, J.C. (1876) *The Geology of the Northern Part of the English Lake District*. Memoir of the Geological Survey of the UK, HMSO, London.

Warren, P.T., Harrison, R.K., Wilson, H.E., *et al.* (1970) Tectonic ripples and associated minor structures in the Silurian rocks of Denbighshire, North Wales. *Geological Magazine*, **107**, 51–60.

Watson, J.V. (1984) The ending of the Caledonian Orogeny in Scotland. *Journal of the Geological Society of London*, **141**, 193–214.

Webb, B.C. (1972) North–south trending pre-cleavage folds in the Skiddaw Slate Group of the English Lake District. *Nature Physical Science*, **235**, 138–40.

Webb, B.C. (1975) The structure and stratigraphy of the Skiddaw Slates between Buttermere and Newlands, Cumbria, and their relationship to the overlying Volcanic Rocks. Unpublished PhD thesis, University of Sheffield.

Webb, B.C. (1983) Imbricate structure in the Ettrick area, Southern Uplands. *Scottish Journal of Geology*, **19**, 387–400.

Webb, B.C. and Cooper, A.H. (1988) Slump folds and gravity slide structures in a Lower Palaeozoic marginal basin sequence (the Skiddaw Group) NW England. *Journal of Structural Geology*, **10**, 463–72.

Webb, B.C. and Lawrence, D.J.D. (1986) Conical fold terminations in the Bannisdale Slates of the English Lake District. *Journal of Structural Geology*, **8**, 79–86.

Wedd, C.B., Smith, B., and Willis, L.J. (1927) *The geology of the country around Wrexham*. Memoir of the Geological Survey of the UK, HMSO, London.

Weir, J.A. (1968) Structural history of the Silurian rocks of the coast west of Gatehouse, Kirkcudbrightshire. *Scottish Journal of Geology*, **4**, 31–52.

Weir, J.A. (1979) Tectonic contrasts in the Southern Uplands. *Scottish Journal of Geology*, **15**, 169–86.

Whalley, J.S. (1973) Finite strain and texture variation associated with folds in a greywacke sequence at Rhosneigr, Anglesey. Unpublished MSc thesis, Imperial College, London.

White, S.H. and Knipe, R.J. (1978). Microstructure and cleavage development in selected slates. *Contributions to Mineralogy and Petrology*, **66**, 165–74.

Wilkinson, I. (1987). A finite strain study of the Ordovician volcanic rocks of Snowdonia, North Wales and its implications for a regional strain model. In *Sedimentation and Tectonics of the Welsh Basin*. (eds W.R. Fitches and N.H. Woodcock). Geological Journal, 22, 95–105.

Wilkinson, I. (1988) The Deformation of the Ordovician Volcanic Rocks of Snowdonia, North Wales. Unpublished PhD thesis, University of Wales, Aberystwyth.

Wilkinson, I. and Smith, M. (1988) Basement fractures in North Wales: their recognition and control on Caledonian deformation. *Geological Magazine*, **125**, 301–6.

Williams, A. (1975) Plate tectonic and biofacies evolution as factors in Ordovician correlation. In *The Ordovician System: Proceedings of a Palaeontological Association symposium*. (ed. M.G. Bassett) University of Wales Press, Cardiff, and National Museum of Wales, pp. 18–53.

Williams, H. (1922) The igneous rocks of the Capel Curig district, North Wales. *Proceedings of the Liverpool Geological Society*, **13**, 166–206.

Williams, H. (1927) The geology of Snowdon, North Wales. *Quarterly Journal of the Geological Society of London*, **83**, 346–431.

Williams, H. and Hatcher, R.D. (1982) Appalachian suspect terranes. *Memoir of the Geological Society of America*, **158**, 33–53.

Wilson, J.T. (1966) Did the Atlantic close and then re-open? *Nature*, **211**, 676–681.

Wood, A. (1958) Whitsun field meeting at Aberystwyth. *Proceedings of the Geologists' Association*, **69**, 28–31.

Wood, A. and Smith, A.J. (1958) The sedimentation and sedimentary history of the Aberystwyth Grits (Upper Llandoverian). *Quarterly Journal of the Geological Society of London*, **114**, 163–95.

Wood, D.S. (1969) The base and correlation of the Cambrian rocks of North Wales. In *The Precambrian and Lower Palaeozoic Rocks of Wales*. (ed. A. Wood), University of Wales Press, Cardiff, pp. 47–66.

Wood, D.S. (1971) Studies of strain and slaty cleavage in the Caledonides of North-west Europe and the eastern United States. Unpublished PhD thesis, University of Leeds.

References

Wood, D.S. (1973) Patterns and magnitudes of natural strain in rocks. *Philosophical Transactions of the Royal Society of London*, **A274**, 373–82.

Wood, D.S. (1974) Current views of the development of slaty cleavage. *Review of Earth and Planetary Science*, **2**, 369–401.

Wood, D.S. and Oertel, G. (1980) Deformation in the Cambrian Slate belt of Wales. *Journal of Geology*, **88**, 285–308.

Wood, D.S., Oertel, G., Singh, J., and Bennett, H.F. (1976) Strain and anisotropy in rocks. *Philosophical Transactions of the Royal Society of London*, **A283**, 27–42.

Wood, M. and Nicholls, G.D. (1973) Pre-Cambrian stromatolitic limestones from northern Anglesey. *Nature, Physical Science*, **241**, p. 65.

Woodcock, N.H. (1976) Ludlow series slumps and turbidites and the form of the Montgomery Trough, Powys. *Proceedings of the Geologists' Association*, **87**, 169–82.

Woodcock, N.H. (1984a) Early Palaeozoic sedimentation and tectonics in Wales. *Proceedings of the Geologists' Association*, **95**, 323–35.

Woodcock, N.H. (1984b). The Pontesford Lineament, Welsh Borderland. *Journal of the Geological Society of London*, **141**, 1001–14.

Woodcock, N.H. (1987a) Structural geology of the Llandovery Series in the type area, Dyfed, Wales. In *Sedimentation and Tectonics of the Welsh Basin*. (eds W.R. Fitches and N.H. Woodcock), Geological Journal, 22, 199–209.

Woodcock, N.H. (1987b) Kinematics of strike-slip faulting, Builth Inlier, Mid-Wales. *Journal of Structural Geology*, **9**, 353–63.

Woodcock, N.H. (1988) Strike-slip faulting along the Church Stretton Lineament, Old Radnor Inlier, Powys. *Journal of the Geological Society of London*, **145**, 925–33.

Woodcock, N.H. and Gibbons, W. (1988) Is the Welsh Borderland fault System a terrane boundary? *Journal of the Geological Society of London*, **145**, 915–23.

Woodcock, N.H., Awan, M.A., Johnson, T.E. *et al.* (1988) Acadian transpression in Wales during Avalonia/Laurentia convergence. *Tectonics*, 7, 483–95.

Index

Page numbers in italic refer to figures and page numbers in bold refer to tables.

Index

Index

Index

Index

Index